Advances in Friction, Lubrication, Wear and Oxidation in Metals Manufacturing

Advances in Friction, Lubrication, Wear and Oxidation in Metals Manufacturing

Editors

Guanyu Deng
Hongtao Zhu
Anh Kiet Tieu

MDPI • Basel • Beijing • Wuhan • Barcelona • Belgrade • Manchester • Tokyo • Cluj • Tianjin

Editors
Guanyu Deng
School of Mechanical,
Materials, Mechatronic and
Biomedical Engineering
University of Wollongong
Wollongong
Australia

Hongtao Zhu
School of Mechanical,
Materials, Mechatronic and
Biomedical Engineering
University of Wollongong
Wollongong
Australia

Anh Kiet Tieu
School of Mechanical,
Materials, Mechatronic and
Biomedical Engineering
University of Wollongong
Wollongong
Australia

Editorial Office
MDPI
St. Alban-Anlage 66
4052 Basel, Switzerland

This is a reprint of articles from the Special Issue published online in the open access journal *Metals* (ISSN 2075-4701) (available at: www.mdpi.com/journal/metals/special_issues/friction_oxidation).

For citation purposes, cite each article independently as indicated on the article page online and as indicated below:

LastName, A.A.; LastName, B.B.; LastName, C.C. Article Title. *Journal Name* **Year**, *Volume Number*, Page Range.

ISBN 978-3-0365-7343-4 (Hbk)
ISBN 978-3-0365-7342-7 (PDF)

Contents

About the Editors

Guanyu Deng

Guanyu Deng received his Ph.D degree in Mechanical Engineering from the University of Wollongong, Australia, in 2014. He is currently a research fellow and group leader of a research team on "Manufacturing and Tribology of High Entropy Alloys" at the University of Wollongong. His major research interests include advanced manufacturing, tribology and lubrication, mechanical characterizations (wear, fatigue and nanoindentation), advanced computational models (finite element method, molecular dynamics, machine learning and multi-scale model), novel structural materials (steel, aluminum, titanium and high-entropy alloys), texture and residual stress. He has won several prestigious awards from the Australian Academy of Science, Australian Synchrotron, and the Japan Society for the Promotion of Science. He has published more than 100 papers and delivered over 10 keynote or invited talks at international conferences.

Hongtao Zhu

Hongtao Zhu received his Ph.D degree from Northeastern University, China, in 2000. After working at Shanghai Jiao Tong University as a postdoctoral fellow for two years, he joined the University of Wollongong in 2002. Now, he is an Associate Professor at the School of Mechanical, Materials, Mechatronic and Biomedical Engineering, University of Wollongong, Australia. His major research interests include tribology, contact mechanics, and quantum and molecular dynamic simulation. He leads a research group on the "Wheel and Rail Contact System" at the University of Wollongong. He has authored/co-authored 218 papers. He is the first and leading CI for six industry contract projects. He has been awarded five Australian Research Council Discovery Projects and two Linkage Projects.

Anh Kiet Tieu

Anh Kiet Tieu is currently a senior professor and coordinator of the Engineering Mechanics (EM) Centre in the Advanced Manufacturing Institute at the University of Wollongong. He is well known internationally in tribology, rolling technology, and computational and experimental mechanics. He was awarded the position of Fellow of the Australian Academy of Technological Science and Engineering in 2007, and the William Johnson International Gold Medal in 2012 for lifetime achievements in materials processing research and teaching. His research areas cover tribology, lubrication, rolling technology, and computational mechanics. He has won more than AUD 25.5 million in research grants, has published more than 800 refereed papers, and has been listed among the world's 2% best scientists.

metals MDPI

Editorial

Advances in Friction, Lubrication, Wear and Oxidation in Metals Manufacturing

Guanyu Deng *, Hongtao Zhu and Anh Kiet Tieu

School of Mechanical, Materials, Mechatronic and Biomedical Engineering, University of Wollongong, Wollongong, NSW 2522, Australia
* Correspondence: gdeng@uow.edu.au or guanyudeng@gmail.com

1. Introduction

When quickly reviewing the developments of new materials design and fabrication, and engineering and industrial manufacturing, it was found that tribology is a very complicated and highly challenging field that cannot be avoided to improve the manufacturing cost and increase the material service life.

In recent years, metals manufacturing has been undergoing significant transformation through optimizations of conventional manufacturing methods and systems (casting, forging, welding, etc.) and through the promotion of new manufacturing processes and techniques (advanced rolling, additive manufacturing, nanomanufacturing, etc.), due to the emergence of new materials, such as high-entropy alloys (HEAs), and the growing requirements for manufacturing efficiency and product quality. Simultaneously, tribology, which plays an important role in metals manufacturing, has attracted increasing attention and interest in both the industrial and academic communities. Friction, wear and lubrication between materials or mechanical parts in contacts, and oxidation of engineering materials at high temperatures, are of fundamental importance. However, their multidisciplinary natures are usually underlined in previous research. In general, extensive knowledge about mechanical engineering, manufacturing engineering, contact mechanics, materials science, chemistry and physics is required. With the development of both experimental techniques and computer simulation methods, the origin of friction, principles of lubrication, development and performance of novel lubricants, mechanisms of wear and oxidation for metallic materials and composites during manufacturing processes or under service conditions have become assessable at different length and time scales.

The purpose of this Special Issue is to collect research reports aimed to provide an up-to-date overview and progress in both experimental studies and theoretical investigations on several topics in the field of wear and friction as well as high-temperature oxidation of engineering metallic materials or composites and the development and evaluation of novel lubricants for advanced metal manufacturing or machining.

Citation: Deng, G.; Zhu, H.; Tieu, A.K. Advances in Friction, Lubrication, Wear and Oxidation in Metals Manufacturing. *Metals* **2023**, *13*, 505. https://doi.org/10.3390/met13030505

Received: 9 February 2023
Revised: 24 February 2023
Accepted: 1 March 2023
Published: 2 March 2023

2. Contributions

In the present Special Issue in *Metals*, eleven research papers with high scientific quality have been published, covering the research topics of experimental characterization of oxidation performance of tool steel for metal rolling industry, evaluation of friction and wear properties of various materials, development of novel lubricants and advanced simulations relevant to the tribological contacts and metal manufacturing.

As one of the most widely applied industrial metal manufacturing techniques to produce bulk products, the rolling process is generally divided into cold rolling process [1,2] and hot rolling process [3,4], depending on the rolling temperatures. For the industrial hot rolling process, high-temperature oxidation of steel work roll and strip and high-temperature friction and wear contact between steel work roll and strips often cause severe problems and affect the surface quality of products [5–7]. Therefore, excellent wear and oxidation resistances are generally required for work roll materials to extend the industrial service life of a rolling mill and improve the product surface quality. The article published by Hao et al. [8] aimed to study the oxidation performance of two popular work roll materials, namely indefinite chill (IC) and high-speed steel (HSS), considering the practical environment in the steel hot rolling industry. The authors found that the IC roll material exhibited lower oxidation kinetics in dry air but faster oxidation kinetics in humid air than HSS roll material. In addition, they also observed a large difference in the phase constitutions of the oxidation products between two materials, which resulted in different surface roughness that was believed to affect the roll/steel tribological contacts and cause surface defects, such as scratching and sticking.

Wear performance of a material is largely related to the tribological conditions. An article published by Lee et al. [9] aimed to characterize the friction and wear properties of AISI304 stainless steel under dry, water-lubricated and oil-lubricated tribological conditions. They found that both dry and water-lubricated wear conditions fall in the boundary lubrication regime at a low rotation speed. In contrast, the oil-lubricated wear condition is considered to be near or in the mixed lubrication regime and leads to the formation of a large amount of thin and elongated wear debris. In comparison with the wear tests without lubricant (dry) or with water as a lubricant, the AISI304 stainless steel exhibits a much lower specific wear rate when oil is used as lubricant. Regardless of the lubricating conditions, strain-induced martensitic phase transformation has been found in all stainless steel samples. The wear performance of a material can also be affected by its microstructure. Zhang and co-authors [10] studied the influence of aging treatment on the corrosion and wear resistances of a Ni-Cr-Mo-based C276 superalloy. It was found that the presence of precipitates at grain boundaries due to aging resulted in increases in the friction coefficient, the fluctuations in the friction coefficient during steady-state sliding wear process and also the wear track dimensions. The sensitization degree is also increased with the aging treatment time and severe corrosion occurs.

In order to meet the demands of quick industrial development, the design and fabrication of novel materials with high wear resistance are always required. The study reported by Liu et al. [11] is about mechanical and tribological characterizations of a CoCrFeNiMo high-entropy alloy (HEA), which is a new material concept, different from the conventional alloys, and often has five or more principal elements [12–15]. The authors revealed a large influence of Mo content in the HEA on its wear properties, microstructural features and mechanical loading responses. Both the hardness and abrasion wear resistance of the CoCrFeNiMo HEA were found to be improved when its Mo molar ratio was increased, owing to the solid solution strengthening and precipitation strengthening. The main wear mechanism is severe micro-cutting and delamination when the Mo molar ratio is lower than 1 and changes into severe micro-cutting and micro-fatigue when the Mo molar ratio is increased to 1.5. Li et al. [16] fabricated a Ni-Co-P/Si_3N_4 composite coating with excellent wear properties using Al-Si as a substrate via a pulse-current electroplating process. The electroplated Ni-Co-P/Si_3N_4 coating showed minimum abrasion wear, while the Al-Si substrate showed severe adhesion and abrasion wear in both dry and lubricated conditions. Additionally, it was also noted that the Ni-Co-P/Si3N4 composite coating possessed a hardness of about three-times the Al-Si substrate.

Dong et al. [17] carried out a wear test of steel/titanium under water-lubricated conditions, using black phosphorus (BP) nanosheets as novel lubrication additives. They systematically investigated the influence of BP nanosheet size on the friction and wear responses of a Ti6Al4V alloy during wear contact with GCr15 steel, by adding large BP nanosheet (2–4 μm), medium BP nanosheet (0.3–0.5 μm) and BP quantum dots (6–10 nm), respectively, in the ultrapure water. It was found that all three kinds of BP nanosheets resulted in great reductions in the friction coefficient and specific wear rate compared to pure water, and BP quantum dots were the best lubrication additives. According to their interpretations, the main lubrication mechanisms are ascribed to the BP adsorption film and the tribo-chemical reaction film, which prevent direct contact between titanium alloy and steel counterpart and effectively improve both the anti-friction and anti-wear. An article reported by Benedicto et al. [18] aimed to develop novel sustainable and environmentally friendly alternatives to mineral oil water-based cutting fluids to lubricate and cool during the machining of titanium alloys. They added polyol esters in oil-in-water emulsions and studied the effect of esters' molecular structure as well as the interaction with the metal surface on the formation of lubricant film. Their results revealed an improvement in lubricity by 17% and reduction in tool wear rate by 37%; thus, the newly proposed cutting fluid can be very useful for difficult-to-machine materials to improve the tribological performance and prolong the tool life.

Trzepiecinski et al. [19] focused on the sheet manufacturing of a Ti6Al4V alloy, in terms of a single-point incremental forming (SPIF) experiment and using a grease-free dry anti-friction spray of MoS_2 as the lubricant. They determined the optimal input processing parameters during warm SPIF of a Ti6Al4V sheet with an initial thickness of 0.8 mm based on a split-plot I-optimal design to ensure the maximum formable wall angle. It was found that the tool rotational speed and step size had much more influence on the axial force than the feed rate, while the step size had the most influence on the in-plane SPIF force.

In addition to the experimental techniques, advanced simulation methods, including finite element modeling (FEM) [20,21], molecular dynamic (MD) simulation [22,23] and atomistic-continuum coupled multiscale simulation [24,25], also play very critical roles in the research fields of tribology-related metal manufacturing and friction and wear mechanisms of materials. There are research papers in this Special Issue relevant to these advanced simulation techniques. The first study was reported by Tao et al. [26], which was about a numerical simulation and finite element analysis of shape control characteristics of non-oriented silicon steel for a certain UCMW (universal crown mill with work roll shifting) cold rolling mill. According to their statements, the friction condition between the roll and strip deteriorates due to very large rolling force and the negative shift in work roll shows significant control on the roll gap crown and edge drop. The second study was reported by Zheng et al. [27], which was to investigate the lubrication characteristics of C4-alkance with the addition of iron nanoparticles in the boundary friction system based on a nonequilibrium MD model. A change from sliding friction to rolling friction was found because of the presence of nanoparticles, which acted like ball bearings between two contact surfaces. In addition, their MD simulations also revealed a transition from partial lubrication to full lubrication when the number of C4-alkane molecules was increased. The findings are beneficial to the understanding of the friction phenomenon of a simple lubricant containing nanoparticles within a small confinement. The third study was reported by Zhang et al. [28], which proposed a concurrent multiscale simulation strategy based on the coupling of continuum and atomistic models to study the influence of lubricating conditions (amount of lubricants, contact surface roughness and load) on the three-dimensional contact responses of Al single crystals. In comparison with the pure FEM (not able to capture the accurate information at the atomic scale due to incorrect constitutive law) and MD (very time consuming and having both spatial and temporal limitations) simulations, their multiscale model was validated as very effective in the three-dimensional rough contacts in both dry and lubricated conditions, with significantly reduced computational cost and without sacrificing too much accuracy. When load was increased, the contact area was found to decrease, regardless of lubrication or not. Contribution to load bearing from the lubricant molecules was confirmed by the pressure distribution changes for different loads.

3. Conclusions and Outlook

Research papers published in the "Advances in Friction, Lubrication, Wear and Oxidation in Metals Manufacturing" Special Issue contributed to the update of the state of the art concerning studies on friction, wear and oxidation properties of materials, development and performance of novel lubricants and advanced simulation strategies for friction contacts. The quality and variety of the collected articles are addressed to both academic and industrial researchers who are looking for new information that can contribute to the advancement of future research in these highly challenging research fields.

Funding: This research did not receive any external funding.

Acknowledgments: As Guest Editors, we are very happy to report the success of this Special Issue and sincerely grateful to all authors and reviewers for their significant efforts in providing high-quality publications. Sincere gratitude also goes to editors and editorial assistants of *Metals* for their continuous and strong support. In particular, we would like to warmly acknowledge Sunny He for his valuable assistance during the preparation of this volume.

Conflicts of Interest: The authors declare no conflict of interest.

References

1. Su, L.H.; Lu, C.; Deng, G.; Tieu, A.K.; Li, J.T.; Zhu, H.T.; Li, H.J.; Sun, X.D. Investigation of deformation behavior during cold rolling cladding process of four-layer composite aluminium alloys. *Adv. Mater. Res.* **2013**, *651*, 424–429. [CrossRef]
2. Deng, G.Y.; Tieu, A.K.; Si, L.Y.; Su, L.H.; Lu, C.; Wang, H.; Liu, M.; Zhu, H.T.; Liu, X.H. Influence of cold rolling on the deformation behavior and crystallographic orientation development. *Comput. Mater. Sci.* **2014**, *81*, 2–9. [CrossRef]
3. Deng, G.; Zhu, Q.; Tieu, K.; Zhu, H.T.; Reid, M.; Saleh, A.A.; Su, L.H.; Ta, T.D.; Zhang, J.; Lu, C.; et al. Evolution of microstructure, temperature and stress in a high speed steel work roll during hot rolling: Experiment and modelling. *J. Mater. Process. Technol.* **2017**, *240*, 200–208. [CrossRef]
4. Deng, G.; Tieu, A.K.; Su, L.H.; Zhu, H.T.; Zhu, Q.; Zamri, W.F.H.; Kong, C. Characterizing deformation behavior of an oxidized high speed steel: Effects of nanoindentation depth, friction and oxide scale porosity. *Int. J. Mech. Sci.* **2019**, *155*, 267–285. [CrossRef]
5. Hao, L.; Jiang, Z.Y.; Cheng, X.; Zhao, J.; Wei, D.; Jiang, L.; Luo, S.; Luo, M.; Ma, L. Effect of Extreme Pressure Additives on the Deformation Behavior of Oxide Scale during the Hot Rolling of Ferritic Stainless Steel Strips. *Tribol. Trans.* **2015**, *58*, 947–954. [CrossRef]
6. Deng, G.; Zhu, H.T.; Tieu, A.K.; Su, L.H.; Reid, M.; Zhang, L.; Wei, P.T.; Zhao, X.; Wang, H.; Zhang, J.; et al. Theoretical and experimental investigation of thermal and oxidation behaviours of a high speed steel work roll during hot rolling. *Int. J. Mech. Sci.* **2017**, *131–132*, 811–826. [CrossRef]
7. Deng, G.; Tieu, A.K.; Su, L.H.; Zhu, H.T.; Reid, M.; Zhu, Q.; Kong, C. Microstructural study and residual stress measurement of a hot rolling work roll material during isothermal oxidation. *Int. J. Adv. Manuf. Technol.* **2019**, *102*, 2107–2118. [CrossRef]
8. Hao, L.; Li, T.; Xie, Z.; Duan, Q.; Zhang, G. The Oxidation Behaviors of Indefinite Chill Roll and High Speed Steel Materials. *Metals* **2020**, *10*, 1095. [CrossRef]
9. Lee, Y.S.; Yamagishi, S.; Tsuro, M.; Ji, C.; Cho, S.; Kim, Y.; Choi, M. Wear Behaviors of Stainless Steel and Lubrication Effect on Transitions in Lubrication Regimes in Sliding Contact. *Metals* **2021**, *11*, 1854. [CrossRef]
10. Zhang, P.; Zhang, C.; Zhou, X.; Huang, Z. Reductions of Intergranular Corrosion Resistance and Wear Resistance in a Ni-Cr-Mo-Based Superalloy by Aging-Treatment-Induced Precipitation. *Metals* **2021**, *11*, 1329. [CrossRef]
11. Liu, Y.; Xie, Y.; Cui, S.; Yi, Y.; Xing, X.; Wang, X.; Li, W. Effect of Mo Element on the Mechanical Properties and Tribological Responses of CoCrFeNiMox High-Entropy Alloys. *Metals* **2021**, *11*, 486. [CrossRef]
12. Yeh, J.W.; Chen, S.K.; Lin, S.J.; Gan, J.Y.; Chin, T.S.; Shun, T.T.; Tsau, C.H.; Chang, S.Y. Nanostructured High-Entropy Alloys with Multiple Principal Elements: Novel Alloy Design Concepts and Outcomes. *Adv. Eng. Mater.* **2004**, *6*, 299–303. [CrossRef]
13. Cantor, B.; Chang, I.T.H.; Knight, P.; Vincent, A.J.B. Microstructural development in equiatomic multicomponent alloys. *Mater. Sci. Eng. A* **2004**, *375–377*, 213–218. [CrossRef]
14. Deng, G.; Tieu, A.K.; Lan, X.; Su, L.; Wang, L.; Zhu, Q.; Zhu, H. Effects of normal load and velocity on the dry sliding tribological behavior of CoCrFeNiMo0.2 high entropy alloy. *Tribol. Int.* **2020**, *144*, 106116. [CrossRef]
15. Deng, G.; Tieu, A.K.; Su, L.; Wang, P.; Wang, L.; Lan, X.; Cui, S.; Zhu, H. Investigation into reciprocating dry sliding friction and wear properties of bulk CoCrFeNiMo high entropy alloys fabricated by spark plasma sintering and subsequent cold rolling processes: Role of Mo element concentration. *Wear* **2020**, *460–461*, 203440. [CrossRef]
16. Li, Z.; Ma, F.; Li, D.; Wan, S.; Yi, G.; Geng, G.; Guo, L. Enhanced Mechanical and Tribological Capabilities of a Silicon Aluminum Alloy with an Electroplated Ni-Co-P/Si3N4 Composite Coating. *Metals* **2022**, *12*, 120. [CrossRef]
17. Dong, S.; Wang, W.; Gao, Y.; Deng, G. Tribological Properties of Different-Sized Black Phosphorus Nanosheets as Water-Based Lubrication Additives for Steel/Titanium Alloy Wear Contact. *Metals* **2022**, *12*, 288. [CrossRef]
18. Benedicto, E.; Rubio, E.M.; Aubouy, L.; Saenz-Nuno, M.A. Formulation of Sustainable Water-Based Cutting Fluids with Polyol Esters for Machining Titanium Alloys. *Metals* **2021**, *11*, 773. [CrossRef]
19. Trzepiecinski, T.; Szpunar, M.; Ostrowski, R. Split-Plot I-Optimal Design Optimisation of Combined Oil-Based and Friction Stir Rotation-Assisted Heating in SPIF of Ti-6Al-4V Titanium Alloy Sheet under Variable Oil Pressure. *Metals* **2022**, *12*, 113. [CrossRef]
20. Deng, G.Y.; Lu, C.; Tieu, A.K.; Su, L.H.; Huynh, N.N.; Liu, X.H. Crystal plasticity investigation of friction effect on texture evolution of Al single crystal during ECAP. *J. Mater. Sci.* **2010**, *45*, 4711–4717. [CrossRef]

21. Liu, M.; Tieu, A.K.; Lu, C.; Zhu, H.; Deng, G. A crystal plasticity study of the effect of friction on the evolution of texture and mechanical behavior in the nano-indentation of an aluminum single crystal. *Comput. Mater. Sci.* **2014**, *81*, 30–38. [CrossRef]
22. Lu, C.; Gao, Y.; Michal, G.; Deng, G.; Huynh, N.N.; Zhu, H.; Liu, X.; Tieu, A.K. Experiment and molecular dynamics simulation of nanoindentation of body centered cubic iron. *J. Nanosci. Nanotechnol.* **2009**, *9*, 7307–7313. [CrossRef] [PubMed]
23. Lu, C.; Gao, Y.; Deng, G.Y.; Michal, G.; Huynh, N.N.; Liu, X.; Tieu, A.K. Atomic-scale anisotropy of nanoscratch behavior of single crystal iron. *Wear* **2009**, *267*, 1961–1966. [CrossRef]
24. Tao, H.; Li, H.; Shao, J.; Zhang, J.; Liu, Y.; You, X. Research on Shape Control Characteristics of Non-oriented Silicon Steel for UCMW Cold Rolling Mill. *Metals* **2020**, *10*, 1066. [CrossRef]
25. Zhang, J.; Tieu, K.; Michal, G.; Zhu, H.T.; Zhang, L.; Deng, G.; Wang, H. A damping boundary condition for atomistic-continuum coupling. *Chin. Phys. B* **2017**, *26*, 068702. [CrossRef]
26. Zhang, J.; Zhang, L.; Tieu, A.K.; Michal, G.; Zhu, H.T.; Deng, G. Finite temperature multiscale simulations for 3D nanoscale contacts. *Appl. Mech. Mater.* **2016**, *846*, 288–293. [CrossRef]
27. Zheng, X.; Su, L.; Deng, G.; Zhang, J.; Zhu, H.; Tieu, A.K. Study on Lubrication Characteristics of C4-Alkane and Nanoparticle during Boundary Friction by Molecular Dynamics Simulation. *Metals* **2021**, *11*, 1464. [CrossRef]
28. Zhang, J.; Su, L.; Wang, Z. Concurrent Multiscale Simulations of Rough Lubricated Contact of Aluminum Single Crystal. *Metals* **2020**, *10*, 965. [CrossRef]

metals

MDPI

Article

The Oxidation Behaviors of Indefinite Chill Roll and High Speed Steel Materials

Liang Hao, Tuanjie Li *, Zhongliang Xie, Qingjuan Duan and Guoyuan Zhang

School of Mechano-electronic Engineering, Xidian University, Xi'an 710071, China; haoliang@xidian.edu.cn (L.H.); zlxie@xidian.edu.cn (Z.X.); qjduan@126.com (Q.D.); gyzhang@xidian.edu.cn (G.Z.)

* Correspondence: tjli@mail.xidian.edu.cn; Tel.: +86-931-2976-688; Fax: +86-931-2976-578

Received: 27 July 2020; Accepted: 11 August 2020; Published: 13 August 2020

Abstract: Indefinite chill (IC) roll and high speed steel (HSS) materials have been widely employed to manufacture work rolls as latter and former stands in hot rolling mills. The oxidation of work rolls is of importance for the surface quality of the rolled workpieces. The isothermal oxidation of the IC and HSS materials was conducted at 650 °C and 700 °C in both dry air and humid air. The isothermal oxidation curves indicate that HSS shows faster kinetics than the IC materials in dry air, whereas the opposite occurred in humid air. The oxide scales of the IC materials after the oxidation in both dry air and humid air are made up of two oxide phases. Two oxide phases were found when the HSS oxidized in the dry air and three oxide phases were found when oxidized in the humid air.

Keywords: indefinite chill roll; high speed steel; oxidation behavior; thermogravimetric analyzer (TGA)

1. Introduction

Hot rolling is one of the primary manufacturing processes in the steel industry, which is highly dependent on the property of work rolls. In the beginning, low-carbon and adamite rolls were applied as standard work rolls. However, the absence of carbides in low-carbon rolls led to a low wear resistance and a frequent replacement of rolls. As for the adamite rolls, a rapid heat crack formation and propagation takes place and requires an immediate roll change and redressing [1–3]. The introduction of high chromium (Hi-Cr) steel, high speed steel (HSS) and indefinite chill (IC) materials has made tremendous improvements in roll performance and campaign length. Hi-Cr rolls with 18% Cr are oxidation-resistant and result in the attachment of rolled materials onto work rolls. Adjustments were made to reduce Cr to less than 7% and add Mo, V, W and C to develop HSS [4–7]. They present high wear resistance and strength but are still difficult to oxidize. Further modifications were made to develop IC roll materials, including a considerable amount of graphite, showing a good thermal and oxidation behavior [8–10]. In addition, the graphite in the IC rolls can retard crack propagation and overcome sticking, since it lubricates the contact between the roll and the rolled strips [11–13]. Nevertheless, the hardness of IC rolls is less than that of HSS, so the wear loss for IC rolls is more rapid. Global experience recommends using HSS rolls primarily in former stands, such as F1–F3, and to use IC rolls in later stands in the hot strip mills [14]. Because of the multiphased microstructures in the HSS rolls, there is no clear consensus regarding the oxidation behavior of each phase at the rolling temperatures [15–17]. For instance, Kim et al. [18] proposed that carbides as well as the martensitic matrix be oxidized in a dry atmosphere, and only use a matrix oxidized in a wet atmosphere. Molinari et al. [19] suggested that the oxidation nucleates at the matrix–carbide interfaces and propagates in the matrix without involving the carbides, because of the high oxidation resistance of carbides. Later, they [5] added that some of the carbides do oxidize but show a different oxidation behavior from the matrix. Particularly, M_7C_3 (Cr-rich) carbides do not oxidize at the rolling temperatures, MC (V-rich) carbides display a faster oxidation and M_2C (Mo-rich) carbides show an intermediate behavior. Zhou et al. [20] found that only the matrix and the MC carbides oxidized but the M_7C_3 carbides hardly oxidized, and the effect of water vapor temperature on the morphology is negligible at each oxidation temperature [21]. Garza et al. [22] thought that M_6C (Mo-rich) and M_7C_3 do not oxidize in humid atmospheres but MC carbides show a significant oxidation in a dry atmosphere.

During the hot rolling, the heat conduction from hot strips, deformation heat, and friction heat can result in a 700 °C surface temperature of work rolls in a very short contact time (10^{-2}–10^{-3}), which can then be cooled to room temperature by spraying water in the rest rime of each cycle (2–10^{-1}) [23]. The accumulated time of work rolls in high temperature is less than 30 min in periods of roll replacement. Furthermore, the oxide scales formed on the roll surface peels off due to friction and thermal fatigue [24–29]. The purpose of this paper is to investigate and compare the oxidation behavior of the IC and the HSS roll materials by a combined study of the oxidation kinetics, surface morphology and cross sections of the oxide scales formed in both dry and humid atmospheres.

2. Experimental

The IC roll and HSS materials were cut from the shell part of work rolls, and their chemical compositions are listed in Table 1. Samples were machined to the dimensions of 15 × 10 × 1 mm^3 with a 0.5 mm side length square hole near the top edge of the sample. Two broad surfaces were polished up to a 1 um diamond suspension and the rest surfaces were ground with 1200 grit sand paper. Then, the samples were ultrasonically cleaned in acetone and rinsed with alcohol.

Table 1. Chemical compositions of the indefinite chill (IC) and high speed steel (HSS) materials (wt.%).

Material	Fe	C	Ni	Cr	V	Mo	Si	Mn
IC	Balance	3.35	4.54	1.85	-	0.49	0.77	0.89
HSS	Balance	2.1	-	4.5	4.4	4.9	0.62	0.55

The isothermal oxidation was carried out on a SETSYS Evolution S60/58507(SETARAM Inc., Marseille, France), which is equipped with a vertical hang-down SETSYS balance. The oxidation tests were conducted under two atmospheres, 20% water vapor and dry air. The isothermal oxidation was chosen at 650 and 700 °C for 30 min, respectively [30]. The experimental details are given as follows: (a) the sample was suspended in the furnace chamber, and then the furnace chamber was vacuumed and flushed with Ar until the pressure reached 1 atm; (b) the sample was heated to the testing temperatures at a rate of 25 °C/min and held for 5 min once the testing temperature was reached; (c) the oxidizing gas was introduced and the isothermal oxidation processed for 30min; (d) the isothermal oxidation was terminated by replacing the oxidation gas with Ar to prevent further oxidation and the sample was cooled to room temperature at a rate of 30 °C/min. The mass variations were recorded in this process. The air containing 20% water vapor was obtained by setting the temperature of the water tank at 60 °C [31].

The microstructure analysis was employed on a JSM scanning electron microscope (SEM) equipped with an energy dispersive spectrometer (EDS) (JEOL Ltd., Tokyo, Japan). The morphology and phases of the oxide scales were studied both on the exposure surfaces and the cross sections. In addition, the oxidation kinetics curves were obtained by plotting the mass gain change per surface against the oxidation time.

3. Results

3.1. Microstructure of IC Roll and HSS Materials

Figure 1 shows the backscattered electron (BSE) images of the polished IC and HSS materials prior to the tests, respectively. Three different phases can be identified in the IC material (Figure 1a), and they are graphite, (Fe, Cr)$_3$C (cementite), and martensite, respectively [32]. The HSS (Figure 1b) presents different microstructure characteristics. Three different carbides can be clearly distinguished by their contrasts, morphologies and EDS analysis. The dark areas are V-rich MC carbides, the grey regions are Cr-Fe-rich M_7C_3 carbides, and the white long zones are Mo-rich M_6C carbides [7,22,33]. These microstructure characteristics enable the IC and HSS materials to present special performances.

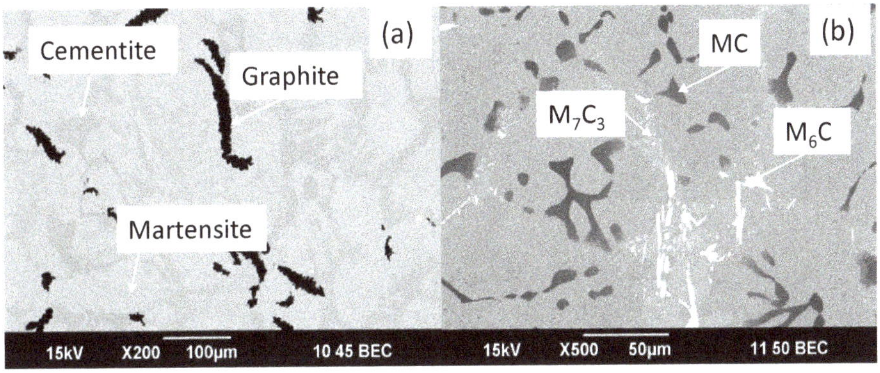

Figure 1. (**a**) Backscattered electron (BSE) images of the polished indefinite chill (IC) and (**b**) high speed steel (HSS) materials prior to the tests.

3.2. Oxidation Kinetics

The isothermal oxidation curves of the IC and HSS materials in both dry and humid air are shown in Figure 2. It is evident that both the temperature and oxidizing atmospheres obviously influence the mass gain of the materials. In dry air (Figure 2a), the mass gain of the HSS is greater than that of the IC rolls at corresponding temperatures, and reveals a parabolic law, while the mass gain of the IC rolls approximately demonstrates a linear law. In addition, the presence of water vapor (Figure 2b) accelerates the mass gain of the IC rolls more than that of the HSS, whereas the oxidation kinetics are little changed by the water vapor.

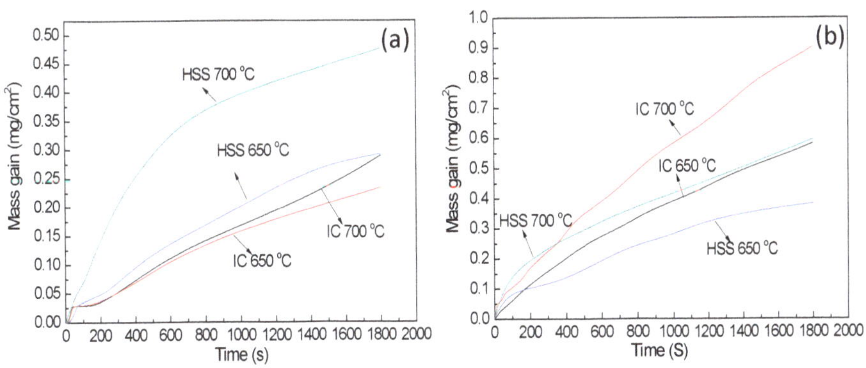

Figure 2. Isothermal oxidation curves of the IC and HSS materials in both (**a**) dry and (**b**) humid air.

3.3. Surface Morphologies

3.3.1. Oxidized in the Dry Air

Figure 3 shows the SEM images of the IC roll surface morphologies oxidized in the dry air at 650 °C and 700 °C. At 650 °C (Figure 3a), the martensite seems to oxidize faster than that of the cementite, and the oxide scales on the martensite protruded out, since the cementite contains higher Cr content and is more oxidation-resistant than the martensite. The extension of the oxide scales nearly covers the graphite areas. When the oxidation processed at 700 °C (Figure 3b), the oxide scales approximately covered the entire surface and obscured the initial boundaries between the cementite and martensite. The cracks were observed in some zones (circled area).

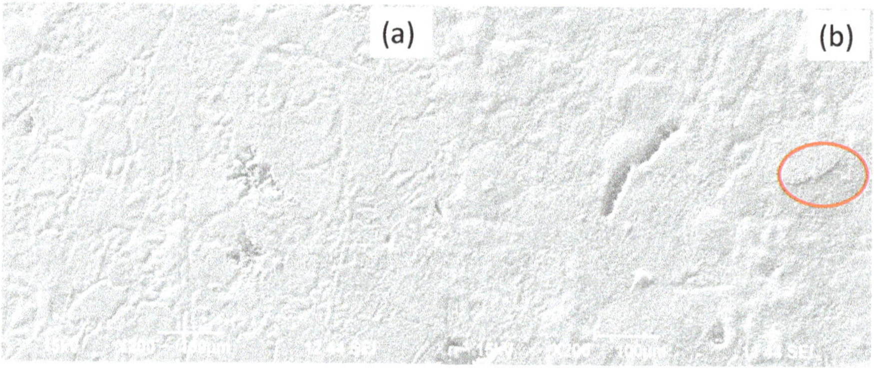

Figure 3. SEM images of the IC roll surface morphologies oxidized in dry air at (**a**) 650 °C and (**b**) 700 °C.

Similar phenomena were observed for the HSS oxidized at 650 °C and 700 °C in the dry air. SEM X-ray maps of the HSS oxidized at 700 °C in the dry air are presented in Figure 4, from which it can be seen that the areas of low O Ka intensity correspond to the areas of the carbides while the areas with a high O Ka intensity are of the matrix. The oxidation of MC carbides extends laterally, whereas the M_7C_3 carbides are still not totally covered by the oxides. Therefore, the matrix oxidized more readily than that of the carbides, and the M_7C_3 carbides are more resistant to oxidization than MC carbides.

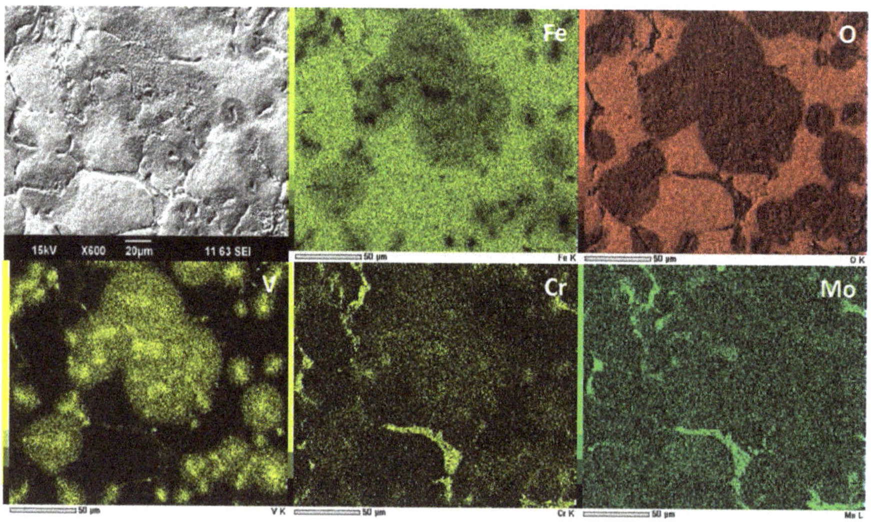

Figure 4. SEM X-ray maps of the HSS oxidized at 700 °C in dry air.

3.3.2. Oxidized in the Humid Air

Figure 5 shows SEM images of the IC roll surface morphologies oxidized in the humid air at 650 °C and 700 °C in the humid air. The IC samples' surfaces were more rough when they oxidized in the humid air than when they oxidized in the dry air. The water vapor accelerates the oxidation for both the cementite and martensite. In addition, the oxide scales formed in the humid air adhered firmly to the matrix without the presence of cracks in the oxide scales. The graphite was nearly covered by the extension of the oxide scales nearby.

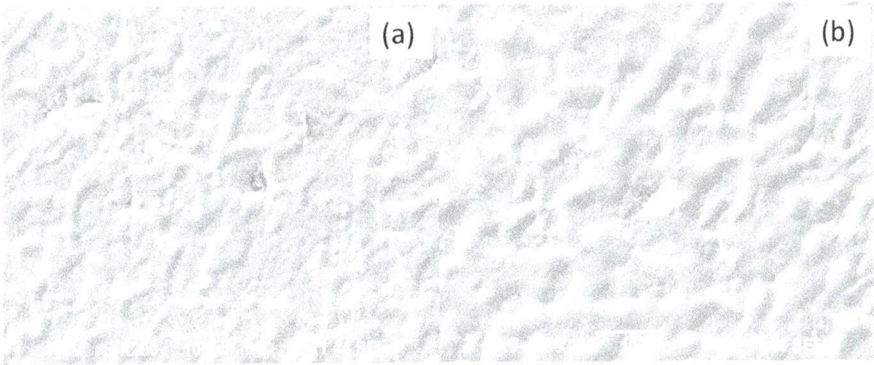

Figure 5. SEM images of the IC surface morphologies oxidized in humid air at (**a**) 650 °C and (**b**) 700 °C.

The HSS samples oxidized faster at 650 °C and 700 °C in the humid air. SEM X-ray maps of the HSS oxidized at 700 °C in the humid air are presented in Figure 6. The similar phenomena were observed as its counterpart in the dry air: the areas with low O Ka intensities correspond to the areas of the carbides, while the areas with high O Ka intensities are of the matrix; MC carbides are oxidized to extend the areas nearby while M_7C_3 carbides are still not entirely covered by the oxide scales.

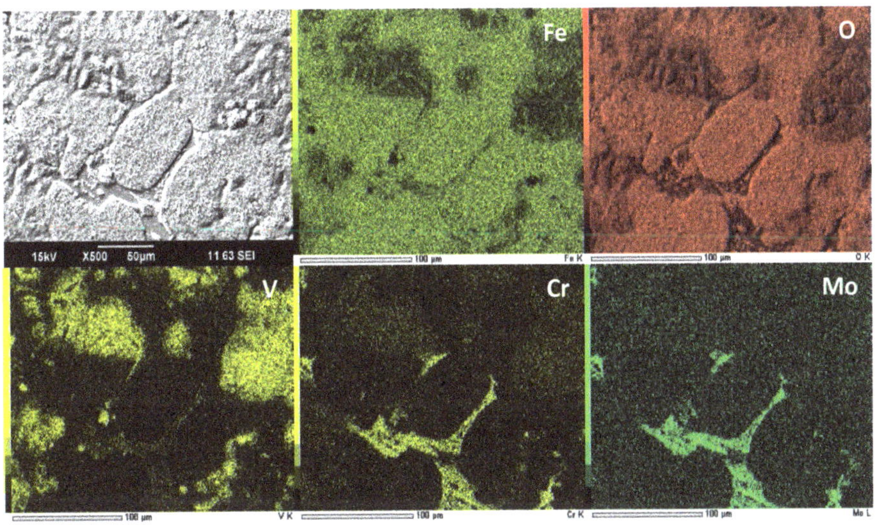

Figure 6. SEM X-ray maps of the HSS oxidized at 700 °C in humid air.

4. Cross Section Examination

The isothermal oxidation curves (Figure 2) demonstrate that the HSS oxidized faster than the IC in the dry air, whereas the IC rolls showed higher kinetics than their HSS counterparts in the humid air. In addition, the oxidation kinetics of the IC and HSS reveal a linear law and parabolic law, respectively.

Figure 7 shows SEM images of the IC cross section of the matrix regions and the graphite areas oxidized at 700 °C in both dry and humid air. As the IC rolls oxidized in the dry air (Figure 7a,b), two relatively thin oxide scales can be observed, namely a thin Cr-rich $(FeCr)_3O_4$ layer next to the substrate, and an outer layer of hematite (Fe_2O_3) [31]. The oxidation of graphite is known as decarburization and left an amount of cavities, which were filled by the extension of the oxide scales. When the IC rolls oxidized in the humid air, the thickness of the oxide scales were increased from 2.26 um in the dry air to about 22.88 um in the humid air. SEM X-ray maps of the IC rolls oxidized at 700 °C in the humid air (Figure 8) reveal that both Si and Ni are present only in the substrate below the inner oxide scales, while the Cr intensifies in the inner oxide.

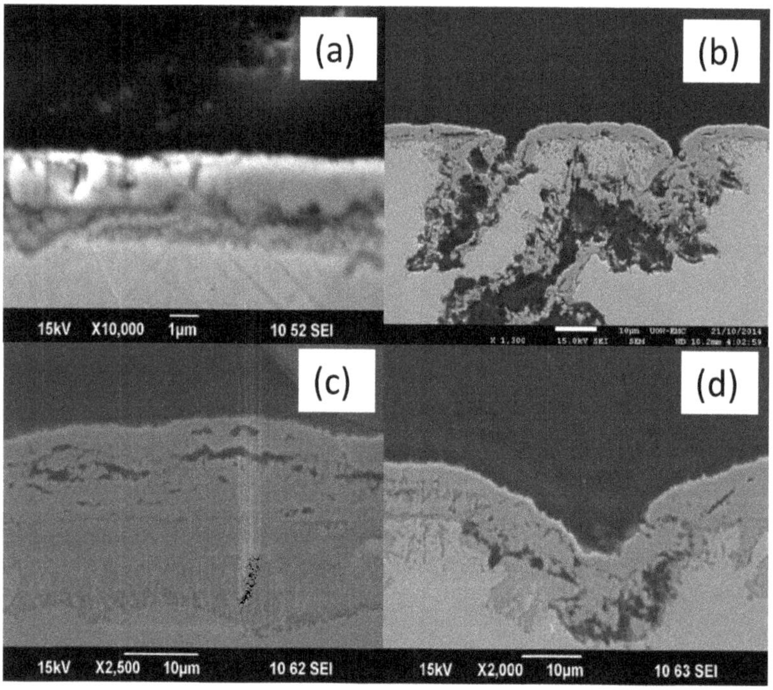

Figure 7. SEM images of the IC roll cross section oxidized at 700 °C: (**a**) the matrix regions and (**b**) the graphite areas in dry air, and (**c**) the matrix regions and (**d**) the graphite areas in humid air.

Figure 9 shows the cross sections and corresponding EDS line scanning analysis of the HSS oxidized at 650 °C in both dry air and humid air. The thickness of the oxide scales was insignificantly increased from 2.54 um in the dry air to 4.56 um in the humid air, which was far less obvious than the thickness increase in the IC roll oxide scales. Two oxide phases can be observed from the EDS line scanning analysis (Figure 9a)—a thin Cr-rich $(Fe,Cr)_3O_4$ next to the substrate and a thick oxide layer of Fe_2O_3 at the scale–gas interface, while three oxide phases, Cr-rich $(Fe,Cr)_3O_4$, magnetite and hematite, can be identified from Figure 9b. Garza et al. [22,34,35] also reported the two oxide phases formed on the HSS oxidized in the dry air, and three oxide phases in the humid air. In addition, the oxide scales tightly adhere to the matrix and no voids are found at the matrix interface.

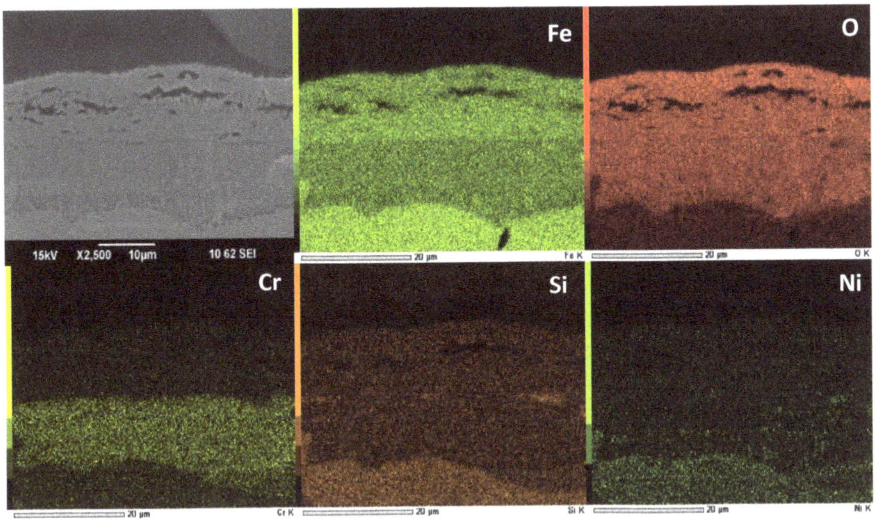

Figure 8. SEM X-ray maps of the IC oxidized at 700 °C in humid air.

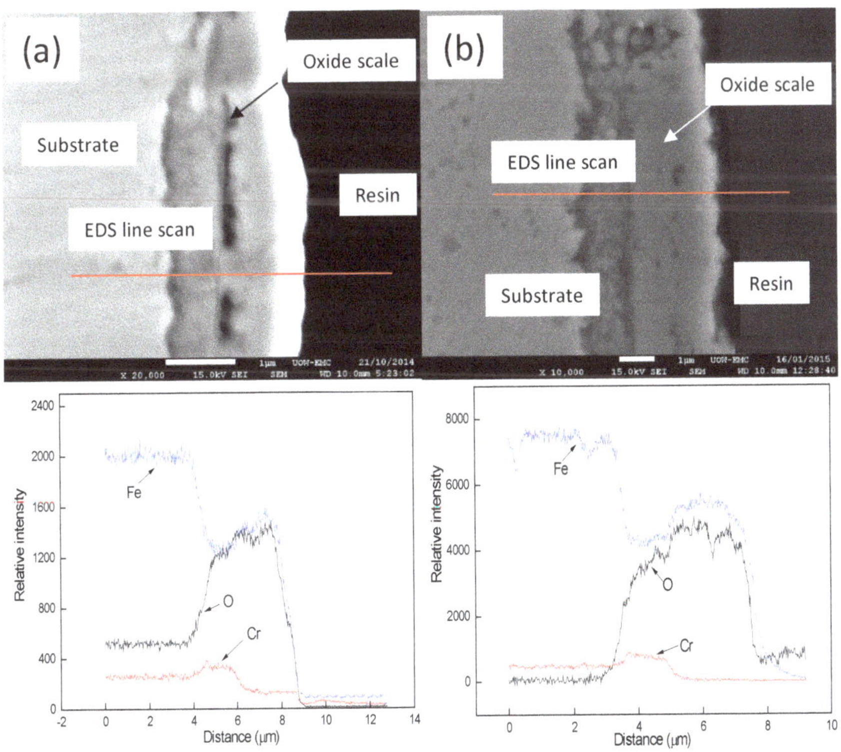

Figure 9. Cross sections and corresponding energy dispersive spectrometer (EDS) line scanning analysis of the HSS oxidized at 650 °C in (**a**) dry air and (**b**) humid air.

The V-rich MC carbides account for the largest fraction of all carbides and a faster oxidation was observed. Figure 10 presents SEM X-ray maps of the cross section for the MC oxidized at 700 °C in the humid air. It is evident that the vanadium oxides extend laterally, which is also confirmed by their surface morphologies in Figures 4 and 6. The possible oxidation process of vanadium-rich carbides is expressed as follows:

$$VC(s) + O_2(g) = VO(s) + CO(g) \quad (1)$$

$$2VC(s) + 3.5O_2(g) = V_2O_5(s) + 2CO(g) \quad (2)$$

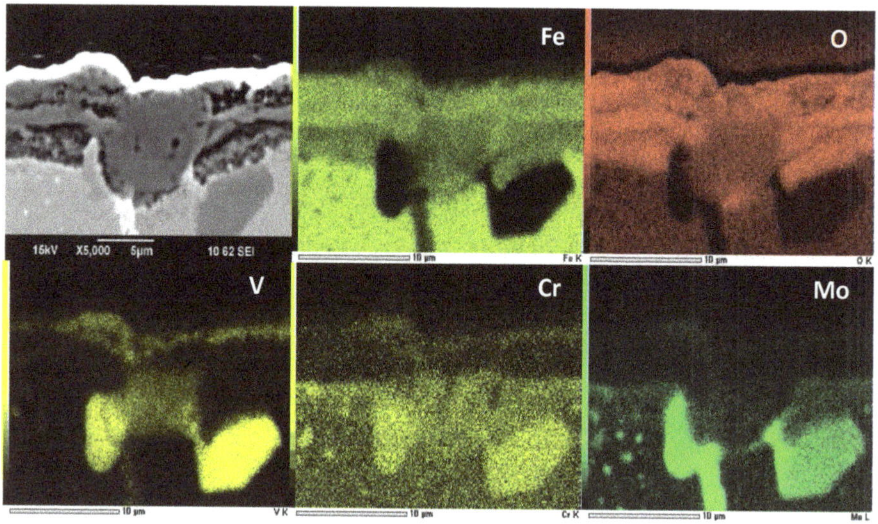

Figure 10. SEM X-ray maps of the cross section for the MC oxidized at 700 °C in humid air.

Some researchers [36,37] proposed that the Gibbs free energy involved in the oxidation of vanadium carbide shows that the formation of V_2O_5 ($\Delta G = -1357$ kJ mol^{-1}, 650 °C) is more favored than that of VO ($\Delta G = -457$ kJ mol^{-1}, 650 °C). Therefore, their conclusion was that the oxidation product of vanadium carbide is V_2O_5. However, the values of ΔG are expressed as kJ mol^{-1} O_2 so that the stabilities of various oxides may be compared directly [38]. In this case, our study indicates that the formation of VO ($\Delta G = -443$ kJ mol^{-1} O_2, 700 °C) is more preferential than that of V_2O_5 ($\Delta G = -383$ kJ mol^{-1} O_2, 700 °C).

5. Conclusions

This isothermal oxidation investigation of the IC and HSS materials has been carried out on a thermogravimetric analyzer (TGA) at 650 °C and 700 °C for 30 min in both dry air and humid air. The following conclusions can be obtained:

1. The HSS exhibits faster oxidation kinetics than the IC in the dry air, whereas the IC rolls oxidized more easily than their HSS counterparts in the humid air. The oxidation kinetics of the IC and HSS materials reveal a linear law and parabolic law, respectively. The oxidation of the HSS suggests that the diffusion of metal cations or oxygen anions is the rate controlling step.
2. The oxide scales of the IC after the oxidation in the dry air and humid air are comprised of two oxide phases: an inner thin Cr-rich $(Fe, Cr)_3O_4$ adjacent to the matrix and an outer Fe_2O_3 at the oxide–gas interface. The oxidation of graphite was faster in the humid air than in the dry air, and denuded cavities were nearly filled by the extension of the oxide scales.

3. Two oxide phases, $(Fe, Cr)_3O_4$, next to the matrix and an outer layer of hematite, were found when the HSS oxidized in the dry air, while three oxide phases, $(Fe, Cr)_3O_4$, magnetite and hematite, were identified in the humid air. The oxidation of V-rich MC carbides was observed in all tested cased. The Gibbs free energy for the oxidation of vanadium carbides indicates that the formation of VO is more preferential than V_2O_5.

Author Contributions: Conceptualization, G.Z.; Formal analysis, Q.D.; Investigation, Z.X.; Supervision, T.L.; Writing—original draft, L.H. All authors have read and agreed to the published version of the manuscript.

Funding: This research was funded by Natural Science Foundation of China (grant number: 51904217) and Natural Science Foundation of Shaanxi Province (grant number: 2020JQ-294).

Acknowledgments: The authors are grateful for the financial support from the National Natural Science Foundation of China (Grant No. 51904217) and Natural Science Foundation of Shaanxi Province (Grant No. 2020JQ-294).

Conflicts of Interest: The authors declare no conflict of interest.

References

1. Ginzburg, V.B. *Flat-Rolled Steel Processes: Advanced Technologies*; Taylor & Francis Group: Abingdon, UK, 2009.
2. Hu, Y.-B.; Zhang, L.; Cheng, C.-Q.; Zhao, P.-T.; Guo, G.-P.; Zhao, J. Oxidation Behavior of the Nickel-Based Superalloy DZ125 at 980 °C. *Acta Metall. Sin. (Engl. Lett.)* **2017**, *30*, 857–862. [CrossRef]
3. Li, X.; Shu, J.; Chen, L.; Bi, H. Effect of Cerium on High-Temperature Oxidation Resistance of 00Cr17NbTi Ferritic Stainless Steel. *Acta Metall. Sin. (Engl. Lett.)* **2014**, *27*, 501–507. [CrossRef]
4. Hao, L.; Jiang, Z.; Wei, D.; Zhao, Y.; Zhao, J.; Luo, M.; Ma, L.; Luo, S.; Jiang, L. Effect of extreme pressure agents on the anti-scratch behaviour of high-speed steel material. *Tribol. Int.* **2015**, *81*, 19–28. [CrossRef]
5. Deng, G.; Tieu, A.; Su, L.; Zhu, H.; Zhu, Q.; Zamri, W.; Kong, C. Characterizing deformation behavior of an oxidized high speed steel: Effect of nanoindentation depth, friction and oxide scale porosity. *Int. J. Mech. Sci.* **2019**, *155*, 267–285. [CrossRef]
6. Fu, H.G.; Qu, Y.H.; Xing, J.D.; Zhi, X.H.; Jiang, Z.Q.; Zhang, Y. Investigations on Heat Treatment of a High-Speed Steel Roll. *J. Mater. Eng. Perform.* **2008**, *17*, 535–542. [CrossRef]
7. Delaunois, F.; Stanciu, V.I. MResistance to High-Temperature Oxidation and Wear of Various Ferrous Alloys Used in Rolling Mills. *Metall. Mater. Trans. A* **2018**, *49*, 822–835. [CrossRef]
8. Pérez, F.J.; Martínez, L.; Hierro, M.P.; Gómez, C.; Portela, A.L.; Pucci, G.N.; Duday, D.; Lecomte, B.J.; Greday, Y. Corrosion behaviour of different hot rolled steels. *Corros. Sci.* **2006**, *48*, 472–480. [CrossRef]
9. Xie, Z.; Shen, N.; Zhu, W.D.; Tian, W.C.; Hao, L. Theoretical and experimental investigation on the influences of misalignment on the lubrication performances and lubrication regimes transition of water lubricated bearing. *Mech. Syst. Signal Process.* **2021**, *148*, 107211.
10. Xie, Z.; Liu, H. Experimental research on the interface lubrication regimes transition of water lubricated bearing. *Mech. Syst. Signal Process.* **2020**, *136*, 106522. [CrossRef]
11. Ming, H.-L.; Wang, S.-Y.; Zhang, Z.-M.; Wang, J.-Q.; Han, E.-H.; Ke, W. Oxidation and Electrochemical Behavior of Monolayer-Graphene-Coated Copper in Simulated Primary Water. *Acta Metall. Sin. (Engl. Lett.)* **2016**, *29*, 89–93. [CrossRef]
12. Deng, G.Y.; Tieu, A.K.; Su, L.H.; Zhu, H.T.; Reid, M.; Zhu, Q.; Kong, C. Microstructural study and residual stress measurement of a hot rolling work roll material during isothermal oxidation. *Int. J. Adv. Manuf. Technol.* **2019**, *102*, 2107–2118. [CrossRef]
13. Vergne, C.; Boher, C.; Gras, R.; Levaillant, C. Influence of oxides on friction in hot rolling: Experimental investigations and tribological modelling. *Wear* **2006**, *260*, 957–975. [CrossRef]
14. Andersson, M.; Finnström, R.; Nylén, T. Introduction of enhanced indefinite chill and high speed steel rolls in European hot strip mills. *Ironmak. Steelmak.* **2004**, *31*, 383–386. [CrossRef]
15. Liu, G.; Xu, Y.-L.; Yang, C.-X.; Xiao, X.-S.; Chen, X.-M.; Zhang, X.-K.; Meng, X.-J. Effects of Alloy Elements on Oxidation Resistance and Stress-Rupture Property of P92 Steel. *Acta Metall. Sin. (Engl. Lett.)* **2015**, *28*, 129–138. [CrossRef]
16. Yuan, X.; Yao, Y.; Chen, L. High-temperature Oxidation Behavior of a High Manganese Austenitic Steel Fe–25Mn–3Cr–3Al–0.3C–0.01N. *Acta Metall. Sin. (Engl. Lett.)* **2014**, *27*, 401–406. [CrossRef]

17. Pellizzari, M.; Cescato, D.; de Flora, M.G. Hot friction and wear behaviour of high speed steel and high chromium iron for rolls. *Wear* **2009**, *267*, 467–475. [CrossRef]
18. Kim, H.H.; Lim, J.L.; Lee, J.J. Oxidation behavior of high-speed steels in dry and wet atmospheres. *ISIJ Int.* **2003**, *43*, 1983–1988. [CrossRef]
19. Molinari, A.; Straffelini, G.; Tomasi, A.; Biggi, A.; Corbo, G. Oxidation behaviour of ledeburitic steels for hot rolls. *Mater. Sci. Eng.* **2000**, *A280*, 255–262. [CrossRef]
20. Zhou, L.; Liu, F.; Liu, C.S.; Sun, D.L.; Yao, L.S. High temperature oxidation behavior of high speed steel for hot rolls material. *J. Univ. Sci. Technol. Beijing* **2005**, *12*, 166–171.
21. Zhou, L.; Liu, F.; Liu, C.S.; Sun, D.L. High temperature oxidation behavior of high speed steel for roll in water vapor. *Trans. Mater. Heat Treat.* **2004**, *5*, 134–138.
22. Garza-Montes-de-Oca, N.F.; Colás, R.; Rainforth, W.M. High temperature oxidation of a work roll grade high speed steel. *Oxid. Met.* **2011**, *76*, 451–468. [CrossRef]
23. Kato, O.; Yamamoto, H.; Ataka, M.; Nakajima, K. Mechanisms of surface deterioration of roll for hot strip rolling. *ISIJ Int.* **1992**, *32*, 1216–1220. [CrossRef]
24. Guo, Y.-L.; Jia, L.-N.; Zhang, H.-R.; Kong, B.; Huang, Y.-L.; Zhang, H. Microstructure and High-Temperature Oxidation Behavior of Dy-Doped Nb–Si-Based Alloys. *Acta Metall. Sin. (Engl. Lett.)* **2018**, *31*, 742–752. [CrossRef]
25. Jiang, Q.-C.; Zhao, X.-M.; Qiu, F.; Ma, T.-N.; Zhao, Q.-L. The Relationship Between Oxidation and Thermal Fatigue of Martensitic Hot-Work Die Steels. *Acta Metall. Sin. (Engl. Lett.)* **2018**, *31*, 692–698. [CrossRef]
26. Wang, R.; Mei, H.-J.; Li, R.-S.; Zhang, Q.; Zhang, T.-F.; Wang, Q.-M. Friction and Wear Behavior of AlTiN-Coated Carbide Balls Against SKD11 Hardened Steel at Elevated Temperatures. *Acta Metall. Sin. (Engl. Lett.)* **2018**, *31*, 1073–1083. [CrossRef]
27. Bongiorno, V.; Piccardo, P.; Anelli, S.; Spotorno, R. Influence of Surface Finishing on High-Temperature Oxidation of AISI Type 444 Ferritic Stainless Steel Used in SOFC Stacks. *Acta Metall. Sin. (Engl. Lett.)* **2017**, *30*, 697–711. [CrossRef]
28. Joos, O.; Boher, C.; Vergne, C.; Gaspard, C.; Nylen, T.; Rezaï-Aria, F. Assessment of oxide scales influence on wear damage of HSM work rolls. *Wear* **2007**, *263*, 198–206. [CrossRef]
29. Bombač, D.; Gintalas, M.; Kugler, G.; Terčelj, M. Mechanisms of Oxidation Degradation of Cr12 Roller Steel during Thermal Fatigue Tests. *Metals* **2020**, *10*, 450. [CrossRef]
30. Deng, G.Y.; Zhu, H.T.; Tieu, A.K.; Su, L.H.; Reid, M.; Zhang, L.; Wei, P.T.; Zhao, X.; Wang, H.; Zhang, J.; et al. Theoretical and experimental investigation of thermal and oxidation behaviours of a high speed steel work roll during hot rolling. *Int. J. Mech. Sci.* **2017**, *131*, 811–826. [CrossRef]
31. Hao, L.; Jiang, Z.; Chen, Z.; Wei, D.; Cheng, X.; Zhao, J.; Luo, M.; Ma, L.; Luo, S.; Jiang, L. High temperature oxidation of indefinite chill roll material under dry and humid atmospheres. *Steel Res. Int.* **2015**, *86*, 1–10. [CrossRef]
32. Byun, G.; Oh, S.; Lee, C.; Lee, S. Correlation of microstructure and microfracture mechanism of five work rolls. *Metall. Mater. Trans. A* **1999**, *30*, 234–243. [CrossRef]
33. Hao, L.; Jiang, Z.; Cheng, X.; Zhao, J.; Wei, D.; Jiang, L.; Luo, S.; Luo, M.; Ma, L. Effect of extreme pressure additives on the deformation behavior of oxide scale during the hot rolling of ferritic stainless steel strips. *Tribol. Trans.* **2015**, *58*, 947–954. [CrossRef]
34. Garza-Montes-de-Oca, N.F.; Rainforth, W.M. Wear mechanisms experienced by a work roll grade high speed steel under different environmental conditions. *Wear* **2009**, *267*, 441–448. [CrossRef]
35. Garza-Montes-de-Oca, N.F.; Colás, R.; Rainforth, W.M. Failure Modes of the Oxide Scale Formed on a Work Roll Grade High Speed Steel. *Oxiadtion Met.* **2011**, *76*, 149–160. [CrossRef]
36. Zhu, Q.; Zhu, H.T.; Tieu, A.K.; Kong, C. Three dimensional microstructure study of oxide scale formed on a high-speed steel by means of SEM, FIB and TEM. *Corros. Sci.* **2011**, *53*, 3603–3611. [CrossRef]
37. Zhu, Q.; Zhu, H.T.; Tieu, A.K.; Reid, M.; Zhang, L.C. In-situ investigation of oxidation behaviour in high-speed steel roll material under dry and humid atmospheres. *Corros. Sci.* **2010**, *52*, 2707–2715. [CrossRef]
38. Birks, N.; Meier, G.H.; Pettit, F.S. *Introduction to High Temperature Oxidation of Metals*; Cambridge University Press: Cambridge, UK, 2009.

metals

Article

Wear Behaviors of Stainless Steel and Lubrication Effect on Transitions in Lubrication Regimes in Sliding Contact

Yoon-Seok Lee [1], Shunnosuke Yamagishi [2], Masataka Tsuro [2], Changwook Ji [3], Seungchan Cho [4], Yangdo Kim [5,*] and Moonhee Choi [1,*]

1 Electronic Convergence Division, Korea Institute of Ceramic Engineering & Technology (KICET), Jinju 52851, Korea; yoonseok@kicet.re.kr
2 Graduate School of Natural Science and Technology, Okayama University, Okayama 700-8530, Japan; pj2m44hk@s.okayama-u.ac.jp (S.Y.); po1x8kuv@s.okayama-u.ac.jp (M.T.)
3 Advanced Forming Process R&D Group, Korea Institute of Industrial Technology, Ulsan 44413, Korea; cwji@kitech.re.kr
4 Composites Research Division, Korea Institute of Materials Science, Changwon 51508, Korea; sccho@kims.re.kr
5 Department of Materials Science and Engineering, Pusan National University, Busan 46241, Korea
* Correspondence: yangdo@pusan.ac.kr (Y.K.); moonhee77.choi@kicet.re.kr (M.C.)

Abstract: The wear behavior of AISI304 stainless steel was investigated under dry, water-, and oil-lubricated conditions. A block-on-disk wear test was conducted in this work, since the test conditions could be controlled easily. For oil-lubricated contact, a significant amount of thin and elongated cutting chip-like debris was observed. This is attributed to the high lubricating effect of oil. Strain-induced martensitic (SIM) transformation was observed for all AISI304 blocks after the wear test, while AISI304 consisted of a single γ-austenite phase prior to the wear test. The Stribeck curve and the corresponding lubrication regimes were also considered to explain the wear behaviors and lubrication effect of AISI304. In comparison to the dry or water-lubricated conditions, which fall in the boundary lubrication regime at a low rotation speed, it is considered that the high viscosity of the oil-based lubricant causes the lubrication condition to enter the mixed lubrication regime early at a lower speed, thus reducing the specific wear rate over the 100–300 rpm range.

Keywords: AISI304; martensitic transformation; austenite; lubricant; wear; friction; hydrodynamic lubrication

Citation: Lee, Y.-S.; Yamagishi, S.; Tsuro, M.; Ji, C.; Cho, S.; Kim, Y.; Choi, M. Wear Behaviors of Stainless Steel and Lubrication Effect on Transitions in Lubrication Regimes in Sliding Contact. *Metals* **2021**, *11*, 1854. https://doi.org/10.3390/met11111854

Academic Editors: Angelo Fernando Padilha and Guanyu Deng

Received: 30 September 2021
Accepted: 16 November 2021
Published: 18 November 2021

1. Introduction

Austenitic stainless steels are widely used in industrial fields owing to their superior ductility, high heat resistance, high corrosion resistance, and non-magnetic properties; however, they exhibit weak friction and wear properties [1,2]. When austenitic stainless steels come in contact with other steel-based materials with high metallurgical compatibility, severe wear may occur during the manufacturing processes such as sliding, drilling, and cutting, which, in turn, decrease the metal surface quality [3]; this is because metals with high metallurgical compatibility are highly soluble in their counterpart atom lattices. Therefore, these frictionally incompatible metals cause severe adhesive wear [4]. Several methods have been proposed to enhance the mechanical processability and operating efficiency, such as using lubricants to reduce the wear between contact surfaces under mechanical stress. Therefore, most mechanical machining processes of stainless steel are performed using water-based fluid or mineral oil [5].

In this study, a block-on-disk wear test was conducted owing to its relatively simple operation. In addition, compared to an actual machining test, the test conditions could be controlled easily [6]. Herein, AISI304, which is a widely used austenitic stainless steel, was used for the wear test under water- and oil-based lubricant contact conditions. Mineral oil, which is an oil-based lubricant that is extracted from petroleum, is commonly used in industrial fields. It contains chemicals such as paraffin and naphthene [7]. Although mineral oil lubricants are widely used in various industries, water-based lubricants are also attracting increasing attention [8].

In severe wear, the evolution of subsurface layer is observed along with that of the contact surface, thereby altering the mechanical properties of AISI304. Thus, studying the changes in the metallurgical microstructure is crucial for the applications of stainless steel. In these subsurface layers, strain-induced martensitic (SIM) transformations occur due to frictional stress, causing changes from γ-fcc to ε-hcp and α'-bcc phases [3,9,10]. Although a martensitic structure may increase the surface hardness, it decreases the corrosion resistivity, which leads to stress corrosion cracking. Moreover, wear debris is an important by-product that can be used to identify the machining conditions [11]. For example, wear behavior can be inferred from the amount of debris, size, shape, and composition [12–16]. Thus, the debris produced during the wear test was also sampled and analyzed.

In any industrial process, phenomena such as oxidative, abrasive, adhesive, and fatigue wear commonly occur. Thus, wear processes should be examined to understand the functionality of different parts under the combined wear conditions. Furthermore, to improve the wear resistance of stainless steel and expand its application scope, the metallurgical mechanisms, such as the lubrication effect with subsurface layers, should be investigated in detail, as research investigating wear behaviors based on metallurgy is still lacking. Therefore, research studies should focus on increasing the life cycle of parts, thereby allowing effective metal formation in industrial fields.

2. Experimental Procedure

2.1. Materials

AISI304 rod (C: 0.09, Si: 0.32, Mn: 1.29, P: 0.00, S: 0.15, Ni: 8.37, Cr: 20.32, and Fe: bal. mass%) with a diameter of 10 mm was cut for using as a cylindrical block, as shown in Figure 1a. Furthermore, the AISI304 blocks were subjected to a solution treatment at 1373 K for 3600 s followed by water quenching. The Vickers hardness of the contact surfaces were also measured at a load of 9.8 N after heat treatment, which is the maximum load of the hardness tester (MVK-C, Akashi Co. Ltd., Takahama, Japan), for a holding time of 15 s. Nine different points were measured for each block. The highest and lowest values were discarded, and the hardness of each block was determined using the average of the remaining seven values.

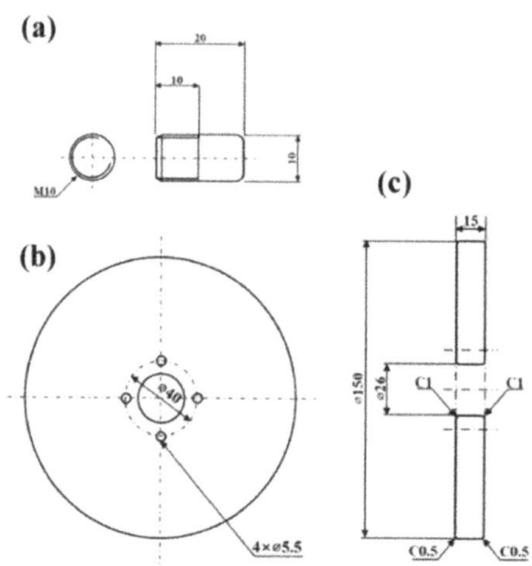

Figure 1. Specifications of (**a**) block, (**b**) disk, and (**c**) rod of experimental setup used for frictional wear test.

Moreover, high speed steel (HSS) was used as a disk and a counter material in this wear-testing system, as shown in Figure 1b,c exhibiting specification of the rod used for fixing the blocks and adding the normal load of 7.3 N to the blocks.

2.2. Block-on-Disk Sliding Wear Test

The block-on-disk wear-testing system was used in this study. The sliding wear is caused by a loaded block contacting a rotating disk. The wear tests were carried out in air at room temperature and constant humidity. The rotation speeds of the disk for each wear test were 100, 200, and 300 rpm, which indicates the sliding speeds of 0.79, 1.57, and 2.36 mm/s, respectively. Then AISI304 blocks were cleaned with acetone and dried in air in order to avoid contamination, before and after each wear test. The weight of each test block, before and after the wear test, was measured using a precise electric balance (0.0001 g). The wear weight loss (W_{loss}) was obtained in grams from the following equation:

$$W_{loss} = W_2 - W_1, \quad (1)$$

where W_1 and W_2 are respectively the weight of a block before and after the wear test. W_{loss} can be converted into the volume loss (V_{loss}) in mm^3 using the following equation:

$$V_{loss} = \frac{W_{loss}}{D_{alloy}} \quad (2)$$

D_{alloy} is the weight per unit volume for the density of AISI304 block (approximately 7.72 g/cm^3) obtained by calculations based on molecular weights. Then, the specific wear rate (W_s) was calculated using the following equation:

$$W_s = \frac{V_{loss}}{F_n L} \quad (3)$$

where F_n is the normal load applied in N and L is the sliding distance in m.

Most of the lubricants which are available in the market are based on mineral oils formulated from petroleum oil. In this study, the petroleum oil (n-butyl stearate $\leq$5, sorbitan oleate $\leq$5, and distillates: bal. mass%) was also used as an oil-based lubricant in the wear-testing system.

2.3. Analysis of Wear Characteristics

The wear debris was sampled after the wear tests and analyzed using scanning electron microscopy (SEM, JIB-4500, JEOL Co. Ltd., Tokyo, Japan) combined with energy-dispersive spectroscopy (EDS). The wear tracks were also examined to understand the wear mechanisms.

Furthermore, some of the AISI304 blocks were sectioned parallel to the sliding direction to examine subsurface layers. SEM based electron back-scattered diffraction (EBSD) analysis was performed to develop a more quantitative view of deformation-induced nucleation of martensite in subsurface layers of the blocks. The cut surfaces were wet polished using SiC water proof papers and were subsequently buff-polished using an Al_2O_3 powder with an average diameter of 0.3 μm and a colloidal SiO_2 solution. The orientation data obtained from the EBSD scans were analyzed using an orientation imaging microscopy (OIM™) analysis software (version 7.2, EDAX Inc., Mahwah, NJ, USA).

3. Results and Discussion

3.1. Wear Behaviors and Characteristics

When stainless steel was treated thermally and it became fully austenitic, the carbides, martensitic structure, and residual stress within the steel were removed, thereby improving its processability and corrosion resistivity. Therefore, to examine the microstructure of the thermally treated test block, its center of the contact surface was analyzed via EBSD prior to the wear test. As shown in Figure 2, the AISI304 sample comprised a single γ-austenite phase after heat treatment.

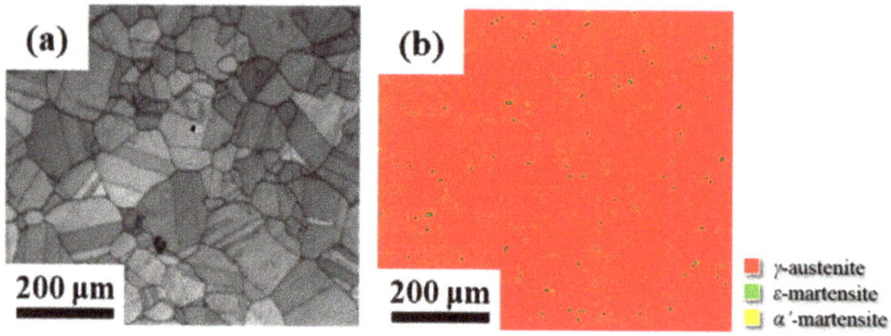

Figure 2. Electron back-scattered diffraction (EBSD) analysis of AISI304 blocks before wear test: (**a**) image quality (IQ) maps and (**b**) corresponding phase maps (PM).

The Vickers hardness of the post-treated AISI304 block was calculated as 150.3 $\pm$ 4.47 HV; in contrast, the hardness value of its counterpart, HSS disk, was 678 HV, which is ~540 HV higher than that of the AISI304 block. W_s for each test condition was calculated using Equations (1)–(3), as shown in Figure 3. At all the rotation speeds, W_s was the highest under dry conditions, followed by those under the water- and oil-lubricated conditions. Typically, under metal-to-metal contact, W_s is proportional to the sliding speed owing to frictional heat and adhesion. This behavior was observed under dry conditions, as shown in Figure 3. However, for the water-and oil-lubricated contacts at 300 rpm and 200 and 300 rpm, respectively, W_s decreased. This is explained in the following sections.

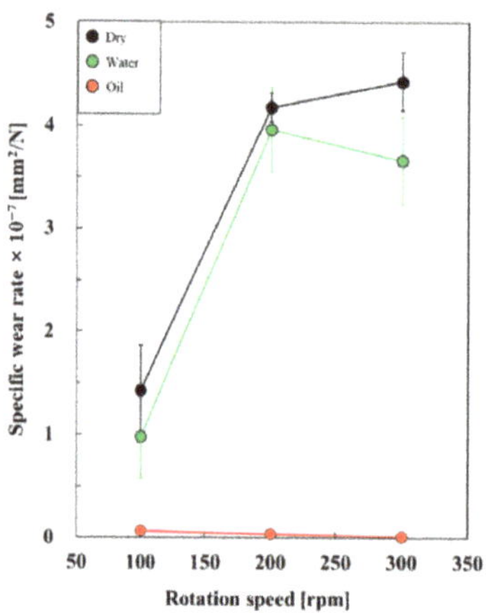

Figure 3. W_s of AISI304 blocks against HSS disk obtained from wear tests under dry, water-, and oil-lubricated conditions.

To understand the wear behavior and the mechanism through the wear tracks on the wear surfaces of the test blocks, SEM images were obtained, as shown in Figure 4. As shown in Figure 4a,b for both low (100 rpm) and high (300 rpm) rotation speeds, the wear tracks of the test blocks under the dry condition showed a significant plastic deformation with smeared surfaces than those under water- and oil-lubricated conditions, thereby indicating strong adhesion forces during wear. In particular, at 300 rpm (Figure 4b), numerous cracks had grown perpendicular to the slide direction, indicating that the fatigue wear causing subsurface cracks is also one of the major wear mechanisms.

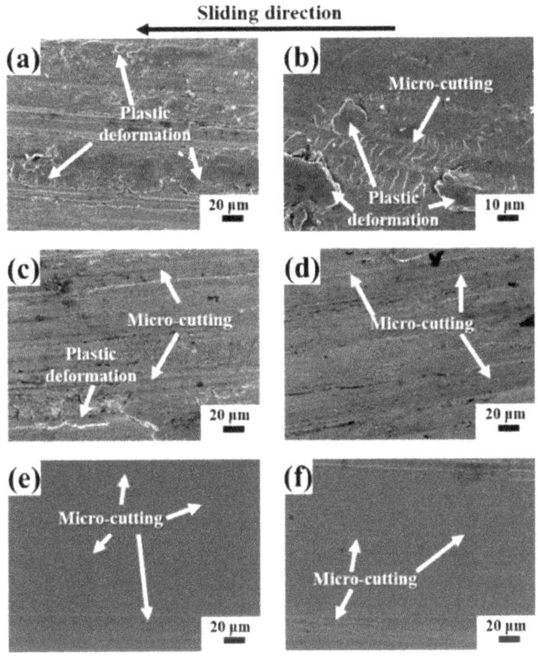

Figure 4. Scanning electron microscopy (SEM) micrographs of wear tracks on AISI304 blocks at rotation speeds of (**a**) 100 and (**b**) 300 rpm, obtained from tests under dry conditions, rotation speeds of (**c**) 100 and (**d**) 300 rpm, obtained from tests under water-lubricated conditions, and rotation speeds of (**e**) 100 and (**f**) 300 rpm, obtained from tests under oil-lubricated conditions.

Furthermore, micro-cutting plays a crucial role in the water-lubricated contact than that in the dry contact, which is a characteristic of abrasive wear. In the case of oil-lubricated contact, regular grooves with micro-cutting are more apparent. Furthermore, a significant decrease in severe plastic deformation was observed than the dry and water-lubricated contacts, regardless of the rotation speed, which confirms the presence of mild wear, indicating a dramatic reduction in the adhesion.

Both the disk and block comprise the same main chemical composition (Fe), demonstrating high metallurgical compatibility. Owing to the high metallurgical compatibility between the disk and block, strong adhesion is prone to occur during wear under the conditions used in this experiment. When water or oil is applied, a lubricating film is formed on the contact surface between the disk and block, decreasing the area of real contact (A_r). Adhesion occurs at the contact surface owing to the interatomic thermal interaction, and it decreases proportionally with A_r. Thus, the decrease in shear stress reduces the plastic deformation of the contact surfaces, and micro-cutting is clearly observed, as shown in Figure 4c–f. In particular, for oil-lubricated contact, the plastic deformation is less severe on the wear tracks, suggesting that the lubricating effect of oil-based lubricant is greater than that of water-based lubricant.

As shown in Figure 5, the dry and water-lubricated contact produced significant amounts of irregular blocky and plate-like debris, clearly indicating fatigue wear [11,17–20]. In the oil-lubricated contact, a significant amount of thin and elongated cutting chip-like debris was observed; this is due to the harder asperities leaving micro-cutting grooves on the contact surface of the mating material that generates cutting chip-like debris. Furthermore, cutting chip-like debris confirms abrasive wear. Typically, while abrasive wear is known to have a higher W_s than other wear mechanisms, the oil-lubricated contact exhibits a low W_s, as shown in Figure 3. This is attributed to the high lubricating effect of oil, which causes much milder abrasive wear, as well as a decrease in the adhesive and fatigue wear. The EDS analysis was also carried out to analyze the wear debris. Strong peaks of chrome (Cr) and nickel (Ni) are detected from debris under dry (Figure 6a,b) and other lubricated conditions, which indicates that they are from the test blocks rather than from the disk.

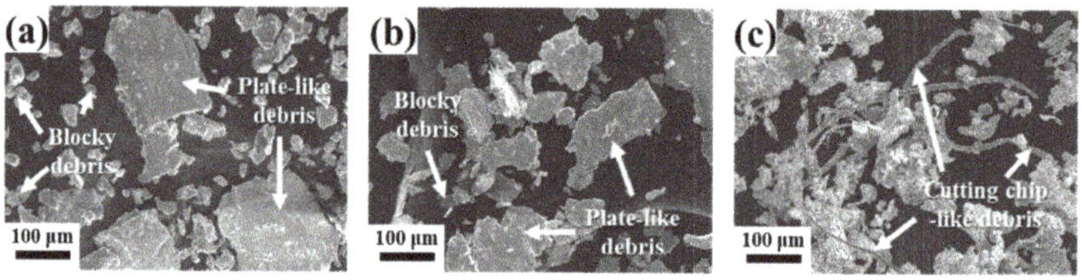

Figure 5. Morphologies of debris collected at the end of wear tests at a rotation speed of 200 rpm and corresponding energy-dispersive X-ray spectroscopy (EDS) analysis; SEM micrographs of wear debris obtained under (**a**) dry, (**b**) water-, and (**c**) oil-lubricated conditions.

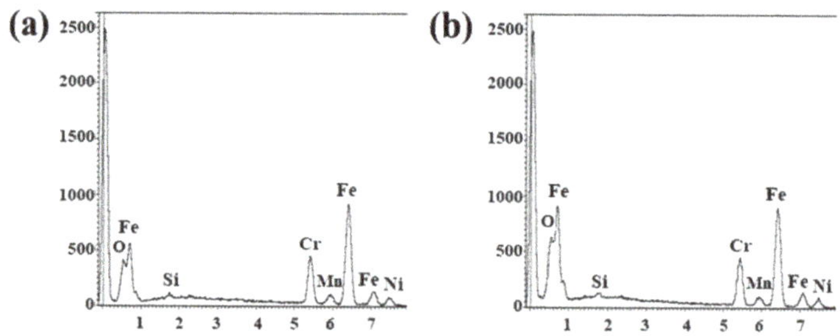

Figure 6. EDS analysis of (**a**) plate-like debris and (**b**) blocky debris collected at the end of wear tests at a rotation speed of 200 rpm under dry contacts.

3.2. *Phase Transformation Behavior*

EBSD analysis was performed to observe the changes in the subsurface layer of the AISI304 block after each wear test, and the phase maps (PM) are shown in Figure 7. Similar to Figure 2, while AISI304 consisted of a single γ-austenite phase prior to the wear test, SIM transformation was observed for all blocks after the wear test.

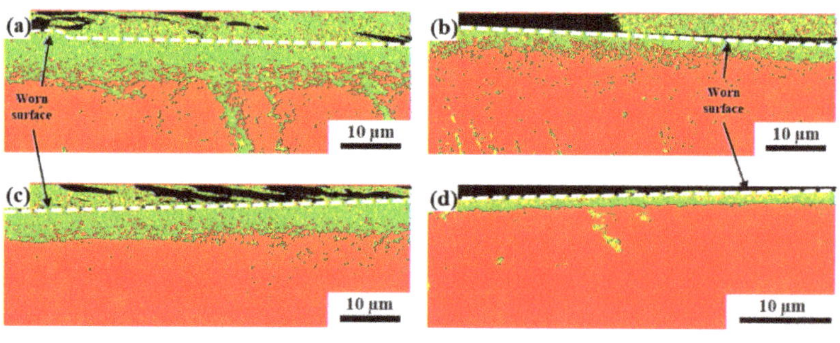

Figure 7. EBSD analysis of cross-sectioned AISI304 blocks after wear tests: phase maps of blocks at rotation speeds of (**a**) 100 and (**b**) 300 rpm obtained from tests under dry contacts, (**c**) 300 rpm obtained from tests under water-lubricated contacts, and (**d**) 300 rpm obtained from tests under oil-lubricated contacts.

It has been reported that high stacking fault energy (SFE) is strongly related to high austenite stability, which retards nucleation and growth SIM [21]. The SFE for austenitic stainless steel can be estimated through empirical equations, based on the chemical composition, using the following equation [22]:

$$\text{SFE (mJ/m}^2) = 32 + 2.4\ (\%\text{Ni}) - 1.2\ (\%\text{Cr}) - 1.2\ (\%\text{Mn}) \quad (4)$$

As SFE value increases, the dominant deformation mode is gradually shifted from ε-martensitic transformation to deformation twinning and then to slip [10]. The calculated SFE value is approximately 26 mJ/m^2 according to Equation (4), in this work. It is reported that the martensitic transformation in AISI304 can also occur at a higher SFE than around 20 mJ/m^2 [23], whereas deformation twinning is favored at SFE more than around 20 mJ/m^2 [24–26]. Furthermore, the SFE is mainly changed by variations in not only chemical composition but also deformation temperature [27–29]. The SFE increases with increasing deformation temperature with same chemical composition [23,28], which can retard nucleation and growth of strain-induced martensite.

In this study, no significant variance in the degree of SIM transformation with rotation speed was observed, excluding dry contact. Theoretically, the friction temperature is a function of the sliding speed. In the case of dry contact at 300 rpm, the frictional heat from high rotation speed caused an increase in the SFE, suppressing the SIM transformation. The high temperature is also responsible for the high adhesion and plastic shearing deformation, resulting in the highest W_s, as shown in Figure 3.

In contrast, in the water-lubricated contact at 300 rpm, the water film at the contact surface can act as a coolant, decreasing the temperature from the frictional heat, resulting in a thicker martensite layer compared to the dry contact at 300 rpm. The oil-lubricated contact had the thinnest martensite layer (~1.5 μm), regardless of the rotation speed.

3.3. Stribeck Curve and Lubrication Region

An empirical criterion, the so-called Stribeck curve, is generally used to identify the three lubrication regimes [30]. Figure 8 shows a schematic of the Stribeck curve and the corresponding lubrication regimes, which are based on the viscosity (η) and sliding speed (V) under the same normal load. The Stribeck curve illustrates the relationship between the lubrication parameter and the basic trend of friction variation. Similar diagrams can easily be found in many books and articles [30,31]. The regime where $h \rightarrow 0$ is the boundary lubrication regime (I), and the regime to its right, where $h \approx R$ (surface roughness), is the mixed lubrication regime (II); herein, hydrodynamic, boundary, and solid lubrication are mixed. Finally, the regime, wherein h >> R, is the hydrodynamic lubrication regime (III).

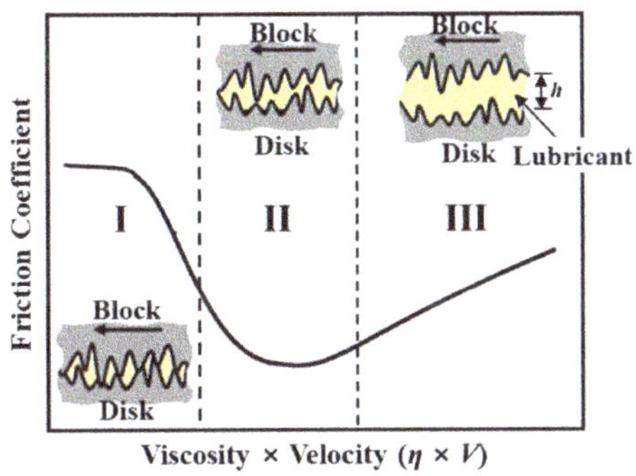

Figure 8. Schematic of the Stribeck curve showing three typical lubrication regimes.

As shown in Figure 3, unlike the dry contact, W_s decrease under the water-lubricated contact at 300 rpm. This may be attributed to the change in the lubrication regime from boundary (I) to the left side in the mixed lubrication regime (II) for the water-lubricated contact, as shown in Figure 8. Thus, at less than 200 rpm, the dry and water-lubricated conditions are both considered to be in the boundary lubrication regime (I), and no significant differences in both the friction coefficient and W_s were observed. The large plastic shear deformation in subsurface layer can lead to accumulation of shear strain and fatigue wear when strong adhesion exists at the contact area [17]. Therefore, in this region (below 200 rpm under dry and water-lubricated conditions), W_s increases with increasing the rotation speed owing to the strong adhesion, although there is no sharp change in friction coefficient. However, above 200 rpm, under the water-lubricated condition, the lubrication regime shifts to the mixed lubrication regime (II), resulting in a significant change in W_s (Figure 3) with a sharp drop in friction coefficient (Figure 8). From the results described above, the amount of martensite at 300 rpm is also larger under water-lubricated condition, compared to that under dry condition. It can be also explained by that shift in regime and a drop in friction coefficient.

Even at the same rotation speed, when the viscosity of the lubricant is sufficiently large, the Stribeck curve moves to the right. For the oil lubricant used herein, even at a low rotation speed (100 rpm), the lubricant is considered to be near or in the mixed lubrication regime (II). In comparison to the dry or water-lubricated conditions, which fall in the boundary lubrication regime (I) at a low speed (100 rpm), the high viscosity of the oil-based lubricant causes the lubrication condition to enter the mixed lubrication regime (II) early at a lower speed, thus reducing W_s over the 100–300 rpm range (Figure 3).

In conclusion, austenitic stainless steel can undergo various severe wear processes such as abrasive, adhesive, and fatigue wear during machining. In particular, when the counterpart material is rich in Fe, strong adhesion may occur, resulting in SIM transformations and subsurface shear deformation. The SIM transformation decreases the corrosion resistance and causes stress corrosion cracking. To prevent this, various water-based and oil-based lubricants are used in drilling and cutting processes. Studies on wear properties, adhesion of contact surfaces, and subsurface deformation under different lubrication conditions can provide fundamental knowledge for machining process applications.

4. Conclusions

In this study, the wear characteristics and behaviors of AISI304 austenitic stainless steel were evaluated at the different rotation speeds, under dry, water-, and oil-lubricated conditions. The primary findings of this study are the following:

1. Owing to the high metallurgical compatibility between the disk and block, strong adhesion is prone to occur during wear. When water or oil is applied, a lubricating film is formed on the contact surface, decreasing A_r and adhesion.
2. In the oil-lubricated contact, a significant amount of thin and elongated cutting chip-like debris was observed. This is attributed to the high lubricating effect of oil, which causes mild abrasive wear.
3. In the water-lubricated contact at 300 rpm, the water film at the contact surface acted as a coolant, decreasing the temperature from the frictional heat, resulting in a thicker martensite layer compared to the dry contact at 300 rpm.
4. At less than 200 rpm, the dry and water-lubricated conditions are both considered to be in the boundary lubrication regime, and no significant differences in both the friction coefficient and W_s were observed. However, above 200 rpm, under the water-lubricated condition, the lubrication regime shifts to the mixed lubrication region, resulting in a significant change in W_s.
5. In comparison to the dry or water-lubricated conditions, which fall in the boundary lubrication regime at a low rotation speed (100 rpm), the high viscosity of the oil-based lubricant causes the lubrication condition to enter the mixed lubrication regime early at a lower speed, thus reducing W_s over the 100–300 rpm range.

Author Contributions: Conceptualization, Y.-S.L.; methodology, Y.-S.L.; formal analysis, S.Y. and M.T.; investigation, S.Y., M.T., C.J. and S.C.; resources, S.Y. and M.T.; data curation, Y.-S.L., S.Y., M.T., C.J. and S.C.; writing—original draft preparation, Y.-S.L.; writing—review and editing, Y.K. and M.C.; visualization, Y.-S.L., S.Y. and M.T.; supervision, Y.K.; project administration, Y.K. and M.C.; Funding acquisition, M.C. All authors have read and agreed to the published version of the manuscript.

Funding: This work was supported by Institute for Information and communications Technology Promotion (IITP) grant funded by the Korea government (MSIP) (No.2021-0-00793).

Institutional Review Board Statement: Not applicable.

Informed Consent Statement: Not applicable.

Data Availability Statement: Not applicable.

Conflicts of Interest: The authors declare no conflict of interest.

References

1. Wei, X.; Hua, M.; Xue, Z.; Gao, Z.; Li, J. Evolution of friction-induced microstructure of SUS 304 meta-stable austenitic stainless steel and its influence on wear behavior. *Wear* **2009**, *267*, 1386–1392. [CrossRef]
2. Lamberson, L.; Mates, S.; Eliasson, V. *Dynamic Behavior of Materials*; Springer: Cham, Switzerland, 2020; Volume 1.
3. McEvily, A.J.; Velazquez, J.L.G. Fatigue crack tip deformation. *Metall. Trans. A* **1992**, *23*, 2211–2221. [CrossRef]
4. Rabinowicz, E. *Friction and Wear of Materials*, 2nd ed.; Wiley: New York, NY, USA, 1995.
5. Junior, A.B.; Diniz, A.E.; Filho, F.T. Tool wear and tool life in end milling of 15–5 PH stainless steel under different cooling and lubrication conditions. *Int. J. Adv. Manuf. Technol.* **2009**, *43*, 756–764. [CrossRef]
6. Xingzhong, Z.; Jiajun, L.; Baoliang, Z.; Hezhou, M.; Zhenbi, L. Wear behavior of Si3N4 ceramic cutting tool material against stainless steel in dry and water-lubricated conditions. *Ceram. Int.* **1999**, *25*, 309–315. [CrossRef]
7. Salih, N.; Salimon, J.; Yousif, E. Synthetic biolubricant basestocks based on environmentally friendly raw materials. *J. King Saud Univ. Sci.* **2012**, *24*, 221–226. [CrossRef]
8. Higdon, C.; Cook, B.; Harringa, J.; Russell, A.; Goldsmith, J.; Qu, J.; Blau, P. Friction and wear mechanisms in AlMgB14-TiB2 nanocoatings. *Wear* **2011**, *271*, 2111–2115. [CrossRef]
9. Fiedler, H.C. The effect of deformation on the martensitic transformation in austenitic stainless steels. *Trans. ASM* **1955**, *47*, 267–290.
10. Choi, J.-Y.; Jin, W. Strain induced martensite formation and its effect on strain hardening behavior in the cold drawn 304 austenitic stainless steels. *Scr. Mater.* **1997**, *36*, 99–104. [CrossRef]
11. Hong, W.; Cai, W.; Wang, S.; Tomovic, M.M. Mechanical wear debris feature, detection, and diagnosis: A review. *Chin. J. Aeronaut.* **2018**, *31*, 867–882. [CrossRef]
12. Seifert, W.W.; Westcott, V.C. A method for the study of wear particles in lubricating oil. *Wear* **1972**, *21*, 27–42. [CrossRef]
13. Dowson, D.; Dalmaz, G.; Childs, T.; Taylor, C.; Godet, M. *Wear Particles: From the Cradle to the Grave*; Elsevier: Amsterdam, The Netherlands, 1992.
14. Roylance, B.; Raadnui, S. The morphological attributes of wear particles—their role in identifying wear mechanisms. *Wear* **1994**, *175*, 115–121. [CrossRef]
15. Eisentraut, K.J.; Newman, R.W.; Saba, C.S.; Kauffman, R.E.; Rhine, W.E. Spectrometric oil analysis. Detecting engine failures before they occur. *Anal. Chem.* **1984**, *56*, 1086A–1094A. [CrossRef]
16. Bharath, V.; Auradi, V.; Nagaral, M.; Boppana, S.B.; Ramesh, S.; Palanikumar, K. Microstructural and Wear Behavior of Al2014-Alumina Composites with Varying Alumina Content. *Trans. Indian Inst. Met.* **2021**, *56*, 1–15. [CrossRef]
17. Suh, N.P. An overview of the delamination theory of wear. *Wear* **1977**, *44*, 1–16. [CrossRef]
18. Wilson, S.; Alpas, A.T. Thermal effects on mild wear transitions in dry sliding of an aluminum alloy. *Wear* **1999**, *225*, 440–449. [CrossRef]
19. Zhang, J.; Alpas, A.T. Delamination wear in ductile materials containing second phase particles. *Mater. Sci. Eng. A* **1993**, *160*, 25–35. [CrossRef]
20. Ashok Raj, J.; Kailas, S.V. Evolution of wear debris morphology during dry sliding of Ti–6Al–4V against SS316L under ambient and vacuum conditions. *Wear* **2020**, *456*, 203378.
21. Talonen, J.; Hänninen, H. Formation of shear bands and strain-induced martensite during plastic deformation of metastable austenitic stainless steels. *Acta Mater.* **2007**, *55*, 6108–6118. [CrossRef]
22. Schramm, R.E.; Reed, R.P. Stacking fault energies of seven commercial austenitic stainless steels. *Metall. Trans. A* **1975**, *6*, 1345. [CrossRef]
23. Noh, J.H.; Lee, S.G.; Park, H.; Kang, N. Effect of Composition on Strain-Induced Martensite Transformation and Tensile Stress-Strain Curve for Austenitic Stainless Steels. *J. Weld. Join.* **2018**, *36*, 28–33. [CrossRef]
24. Olson, G.B.; Cohen, M. A general mechanism of martensitic nucleation: Part I. General concepts and the FCC→HCP transformation. *Metall. Trans. A* **1976**, *7*, 1897–1904.

25. Allain, S.; Chateau, J.P.; Bouaziz, O. A physical model of the twinning-induced plasticity effect in a high manganese austenitic steel. *Mater. Sci. Eng. A* **2004**, *387*, 143–147. [CrossRef]
26. Curtze, S.; Kuokkala, V.T.; Oikari, A.; Talonen, J.; Hänninen, H. Thermodynamic modeling of the stacking fault energy of austenitic steels. *Acta Mater.* **2011**, *59*, 1068–1076. [CrossRef]
27. Saeed-Akbari, A.; Mosecker, L.; Schwedt, A.; Bleck, W. Characterization and prediction of flow behavior in high-manganese twinning induced plasticity steels: Part I. Mechanism maps and work-hardening behavior. *Metall. Mater. Trans. A* **2012**, *43*, 1688–1704. [CrossRef]
28. Walter, M.; Roncery, L.M.; Weber, S.; Leich, L.; Theisen, W. XRD measurement of stacking fault energy of Cr–Ni austenitic steels: Influence of temperature and alloying elements. *J. Mater. Sci.* **2020**, *55*, 13424–13437. [CrossRef]
29. Mosecker, L.; Pierce, D.T.; Schwedt, A.; Beighmohamadi, M.; Mayer, J.; Bleck, W.; Wittig, J.E. Temperature effect on deformation mechanisms and mechanical properties of a high manganese C+ N alloyed austenitic stainless steel. *Mater. Sci. Eng. A* **2015**, *642*, 71–83. [CrossRef]
30. Kadokawa, J. *Ionic Liquids: New Aspects for the Future*; InTech: Rijeka, Croatia, 2013.
31. Williams, J.A. *Engineering Tribology*; Oxford University Press: Oxford, UK, 1994.

metals

MDPI

Article

Reductions of Intergranular Corrosion Resistance and Wear Resistance in a Ni-Cr-Mo-Based Superalloy by Aging-Treatment-Induced Precipitation

Pengyan Zhang [1,2], Chi Zhang [3,*], Xiaoguang Zhou [4] and Zhenyi Huang [1,2]

1 Key Laboratory of Metallurgical Emission Reduction & Resources Recycling, Anhui University of Technology, Ministry of Education, Maanshan 243032, China; zpy5102@163.com (P.Z.); huangzhenyi@ahut.edu.cn (Z.H.)
2 School of Metallurgical Engineering, Anhui University of Technology, Maanshan 243032, China
3 School of Materials Science and Engineering, Dalian University of Technology, Dalian 116024, China
4 State Key Laboratory of Rolling Technology and Automation, Northeastern University, Shenyang 116024, China; zhouxg@ral.neu.edu.cn
* Correspondence: zhangchi@dlut.edu.cn; Tel.: +86-0411-8470-6087

Abstract: Ni-Cr-Mo-based superalloy is widely used as a key component in many critical environments. To ensure that the manufacturing process does not impact the long-term service performance of these components, the aging precipitation behavior at different temperatures and its effect on intergranular corrosion (IGC) resistance and wear resistance of a Ni-Cr-Mo-based C276 superalloy were investigated. The equilibrium phase diagram was calculated first using thermodynamic software to confirm the potential phases. Carbides of M_6C were found to be formed at grain boundaries after aging at 800–850 °C for short-term treatment. The other two phases (μ phase and P phase) indicated in the phase diagram were not observed for the samples after aging treatment up to 15 h. Furthermore, double loop electrochemical potentiokinetic reactivation (DL-EPR) tests were conducted to examine the IGC resistance. The degree of sensitization increased with the aging time and severe corrosion was found to occur at grain boundaries. For the first time, the influence of aging treatment on the wear behavior of this superalloy has been specifically studied. Concerning the hot processing of Ni-Cr-Mo-based C276 superalloy, these results indicate the importance of avoiding high-temperature heat treatment for long periods.

Keywords: Ni-Cr-Mo-based superalloy; aging treatment; precipitation; intergranular corrosion resistance; wear resistance

Citation: Zhang, P.; Zhang, C.; Zhou, X.; Huang, Z. Reductions of Intergranular Corrosion Resistance and Wear Resistance in a Ni-Cr-Mo-Based Superalloy by Aging-Treatment-Induced Precipitation. *Metals* **2021**, *11*, 1329. https://doi.org/10.3390/met11081329

Academic Editor: Belén Díaz Fernández

Received: 6 July 2021
Accepted: 17 August 2021
Published: 23 August 2021

1. Introduction

Superalloys usually preserve excellent high-temperature performance, resulting in them being widely used in many industries with intense conditions, such as aerospace, chemical, nuclear, etc. C276 superalloy is one of the most promising superalloys belonging to the Ni-Cr-Mo-based family, with good corrosion resistance, high temperature strength, and creep resistance [1–4]. This superalloy was first developed by dramatically reducing the content of carbon and optimizing the alloying elements [5]. The outstanding advantage of this superalloy is that it can be used directly in a welding state without severe intergranular corrosion (IGC) attack [6,7].

Due to the excellent properties of C276 superalloy, it is an ideal candidate material of some key components in the nuclear industry [8–10]. The material was designed to be used in temperatures below 650 °C. However, during the manufacture process, C276 superalloy needs to undergo heat treatments at elevated temperatures for a certain period of time in some cases. Recently, a new technology, named hot vacuum bulging technology, was developed to manufacture a rotor can, which is a key component in nuclear reactor coolant pumps, using C276 superalloy [11,12]. The hot vacuum bulging processing is conducted at 800 °C for 2 h to realize exact forming through creep deformation. Creep age forming (CAF) is an effective forming technology for aluminum alloys, and potentially has applications with superalloys [13]. Therefore, the effect of high-temperature aging on performance during the manufacture process (which is much higher than the service temperature) becomes a main concern for C276 superalloy.

C276 superalloy is used in solution state with single austenite phase. There is no need to apply aging treatment to this superalloy to optimize the precipitation of γ' and/or γ'', such as the precipitation-strengthened superalloys [14,15]. Previous investigations into the aging treatment of C276 superalloy focused mainly on testing the stability of performance during long-term service. Liu et al. found that massive block μ phase and M_6C carbides precipitated in the C276 superalloy after aging at 700 °C for more than 360 h, and the precipitates grew along with the aging time [16]. The tensile test at 700 °C indicated that the aged samples maintained the tensile strength with a small decrease in plasticity until 2160 h. The fracture of the tested sample still presented ductile morphology. Akhter et al. tested the effect of aging treatment on the hardness and impact energy of C276 superalloy [17]. The authors found that Mo-rich μ phase precipitated after aging at 850 °C for 48 h, associated with a dramatic decrease in impact energy. Raghavan et al. have characterized the three distinct second phases, μ phase, M_6C carbides, and P phase, in the long-term aged C276 superalloy using transmission electron microscopy (TEM) [5]. The chemical compositions are remarkably close for the three kinds of precipitates. The authors defined the precipitates using a transmission electron microscope.

However, the influence of short-term aging treatment on the performance of this kind of superalloy has been given little attention. There are indications that short-term aging at high temperature can also induce precipitation in other Ni-Cr-Mo-based superalloys [18,19]. A similar phenomenon was also observed in C276 superalloy when it was aging treated at 800 °C for a relatively short time [20]. Furthermore, the nuclear reactor coolant pump rotor serves as a protective barrier for the rotor against cyclic erosion of the coolant. The corrosion resistance and wear resistance of C276 superalloy are key utilized material properties for C276 superalloy. For example, there were clear, synergistic effects between corrosion and wear in this alloy under seawater conditions, leading to corrosion-induced wear and wear-induced-corrosion in tribocorrosion processes [21–23]. According to Hashim and Duraiselvam [24], the tribological properties of C276 superalloy could be improved by laser surface treatment without sacrificing its corrosion properties. Yilbas and Ali also studied the effect of laser treatment on C276 alloy [25]; they found that laser treatment was effective in increasing the surface microhardness and decreasing the friction coefficient. Nevertheless, it is worth noting that no study investigating the effect of aging treatment on the tribological behavior of C276 superalloy has been reported. Our previous paper [20] revealed that small amounts of precipitation resulted in the deterioration of corrosion resistance, but the relationship between precipitation, corrosion behavior, and wear resistance of C276 superalloy is yet to be elucidated. Therefore, the precipitation behavior at different aging treatment temperatures and its influence on the corrosion and wear resistances are highly deserving of an investigation.

The present study is an extension from our previous report [20]. In this work, the equilibrium phase diagram for C276 superalloy was calculated first. Then, based on the thermal dynamic calculation results, aging treatments at two different temperatures were specifically conducted. The aging-treatment-induced precipitations were carefully characterized by SEM and TEM. Moreover, the corrosion resistance, mechanical property, and wear behavior of aging-treated C276 superalloy was evaluated by double-loop electrochemical potentiokinetic reaction (DL-EPR) test, microhardness measurement and ball-on-plate sliding wear test, respectively.

2. Materials and Methods

The tested C276 superalloy is a 0.5 mm thick commercial sheet, which is in solution state with equiaxed austenite grain structure. The chemical composition of the material is: 0.004 C, 0.01 Si, 16.00 Cr, 16.34 Mo, 5.98 Fe, 3.46 W, 0.25 Co, 0.04 Cu, 0.16 V, 0.4 Mn and balance Ni in wt.%. More details are available in Ref. [20]. Based on the chemical compositions, thermodynamic software Thermal-Calc (Thermo-Calc Software, Canonsburg, RA, USA) was used to calculate the equilibrium phase diagram. In order to make a comparison with the previous paper [20], the sheets were aging treated in a vacuum furnace at 800 °C and 850 °C for 1–15 h, followed by water quenching.

The samples used for microstructure analysis were cut from these aging-treated sheets. After mechanical polishing, the samples were light etched in aquaregia (HCl: HNO_3 = 3:1) for 1–3 min and then examined with a field emission scanning electron microscope (Zeiss Supra 55, Zeiss, Germany). For TEM observation, disks 3 mm in diameter and 50 μm in thickness were electrochemically polished at −30 °C in a 5% perchloric acid ethanol solution and examined in JEM-2100F (Jeol Ltd., Japan, JEM-2100F).

The DL-EPR test was performed to investigate the corrosion resistance of the aged samples. Square samples measuring 15 × 15 mm were cut and the sample surfaces were mechanically polished. The conventional solution (0.01 mol/L KSCN and 0.5 mol/L H_2SO_4) was found to not be applicable for distinguishing the current peaks in the electrochemical tests. In this work, we introduced a new solution of 6 mol/L HCl to investigate the IGC resistance of C276 superalloy. A saturated calomel electrode (SCE) is served as the reference electrode. The scan started from the potential of −0.3 V_{SCE} to 0.7 V_{SCE} with a rate of 1 mV/s, then reversed.

The microhardness was examined using the MHV−2.0 Vickers (Shenzhen Shunhua Instrument Equipment Co., Ltd., Shenzhen, China) hardness tester with a load of 0.25 N. The influence of aging treatment on the wear resistance of C276 superalloy was studied in an acceleration condition using a ball-on-plate wear testing rig. The detailed information about the wear testing rig can be found in the study by Li et al. [26]. Prior to the wear tests, C276 superalloy plates were ground and polished to a final surface roughness of less than 100 nm [27]. Al_2O_3 balls with a diameter of 7 mm were selected as the counter-body. Wear tests were performed at room temperature and repeated at least three times for each material. The applied normal load on the ball was 10 N and the sliding speed was 0.92 m/min. The length of the wear stroke was 12 mm and the total reciprocating wear test duration was 30 min. During each test, the friction coefficient was automatically calculated and recorded. The worn surface profile was then examined by an optical 3D surface profiler (Super View W1 type, which is manufactured by Shenzhen Zhongtu Instrument Co., Ltd., Shenzhen, China).

3. Results and Discussion

3.1. Thermodynamic Calculation Results

The equilibrium phase diagram based on the chemical compositions was calculated using thermodynamic software Thermal-Calc. The molar fractions (NP) of phase constituents varying with the temperature is shown in Figure 1. It infers that the μ phase, P phase, and M_6C are supposed to precipitate from the matrix at high temperatures; μ phase can precipitate at temperatures lower than 772 °C, P phase can precipitate in the temperature range of 598–1098 °C, M_6C can form at temperatures lower than 1111 °C, and the content increases with the decreasing temperature. The maximum content for the precipitation of M_6C is 0.14 mol%. The equilibrium chemical compositions of the precipitations are listed in Table 1. The precipitations are mainly composed of Cr, Mo, W, and Ni.

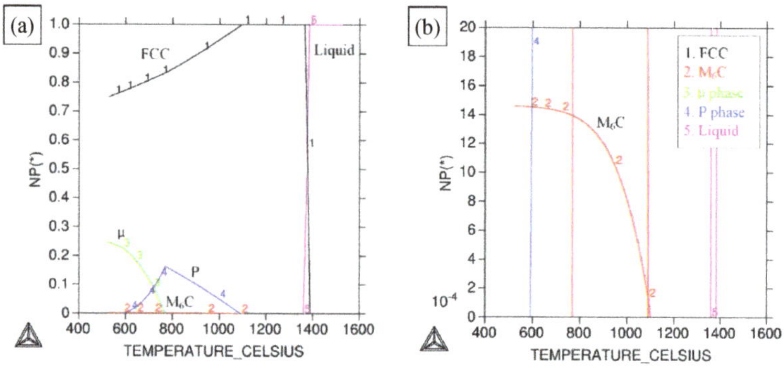

Figure 1. Equilibrium phase diagram (mol%) predicted by Thermo-Calc software for the tested C276 superalloy. * NP means the molar fraction. (**a**) The whole phase diagram; (**b**) magnification of (**a**).

Table 1. The equilibrium chemical composition of the precipitates (wt.%).

	C	Cr	Mo	Fe	W	Co	Ni
P	0	16.00	41.87	1.05	11.5	0.01	29.45
μ	0	16.00	47.32	1.94	5.58	0.01	29.04
M_6C	2.48	9.17	57.9	2.60	5.38	0.007	22.40

It should be noted that the equilibrium phase diagram is calculated based on the thermodynamic analysis. It needs time for the diffusion of alloy atoms to form the precipitates. Three kinds of precipitates have been reported to be formed in Ni-Cr-Mo-based superalloy [5,28]. However, the aging treatments were conducted at 650–900 °C for 1000 h in the investigation of Raghavan et al. [5]. Akhter et al. heated the samples up to 240 h to study the effects of aging treatment on the hardness and impact property [17]. They also mentioned that μ phase cannot form during the aging treatment at 650 °C within 500 h, while M_6C may be produced after 10 h of aging.

3.2. Precipitate Observations

In our previous paper [20], the precipitation behavior of C276 superalloy during aging at 800 °C was studied. In order to understand the influence of aging temperature, the precipitation behavior of C276 superalloy during aging at 850 °C was studied and compared with the results of aging at 800 °C. Figure 2 shows the backscattered electron (BSE) images of the aging-treated samples with different aging temperatures. One can determine that some second phases precipitated along the austenite grain boundaries. By comparison, the number of precipitates increases with the increase in aging time and aging temperature. Furthermore, the morphologies of the precipitates seem very similar for all the samples. The amounts of precipitates were observed to be very low for the samples aged within 4 h. Energy dispersive spectrometer (EDS) analysis was used to detect the compositions of the precipitates, as shown in Figure 3. It can be seen that the precipitates are rich in Mo and W as compared to the base material. The compositions are similar to the thermodynamic calculation results as listed in Table 1. However, it is hard to distinguish the precipitates based on the chemical compositions.

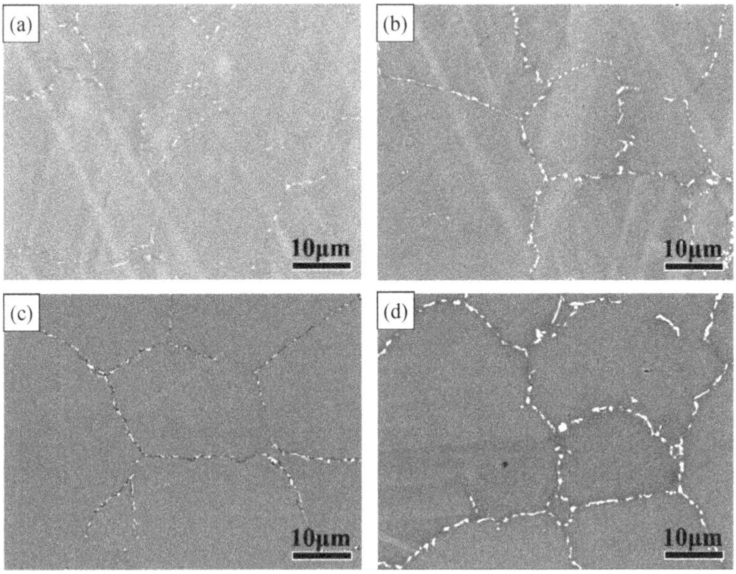

Figure 2. Backscattered electron images of the C276 superalloy after aging treatments: (**a**) 800 °C for 4 h; (**b**) 800 °C for 15 h; (**c**) 850 °C for 4 h; (**d**) 850 °C for 15 h.

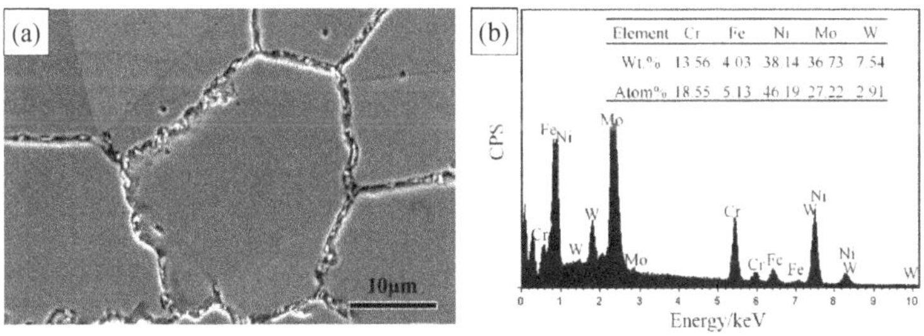

Figure 3. Secondary electron image (**a**) and EDS analysis (**b**) of the C276 superalloy aged at 850 °C for 15 h.

TEM was used to further confirm the precipitates, as shown in Figure 4. The precipitates were formed at grain boundaries with a twinning boundary. The diffraction pattern indicates that the precipitation is a face-centered cubic (FCC) structure with a lattice constant of a = 1.11 nm. M_6C is reported to have a cubic crystal structure with a = 1.08–1.12 nm [5]. μ phase (A_6B_7 type) has a hexagonal close packed structure with a = 0.476–0.479 nm and c = 2.57–2.59 nm, while P has a tetragonal crystal structure [5,16]. Moreover, Raghavan et al. reported that μ phase and M_6C were the main precipitates for C276 superalloy when aging at 650 °C–900 °C up to 1000 h [5]. Based on the phase diagram in Figure 1, μ phase is supposed to precipitate at a temperature below 772 °C. Therefore, the precipitate is supposed to be M_6C, which is consistent with the previous report [20]. No other types of precipitates were observed in TEM analysis. The chemical composition analysis result corresponds to the thermodynamic calculation result in Section 3.1. Therefore, both the thermodynamic calculation and experimental investigation confirm the precipitation of M_6C carbides in the aging-treated C276 superalloy. Increasing the aging temperature to 850 °C does not introduce other precipitates during aging.

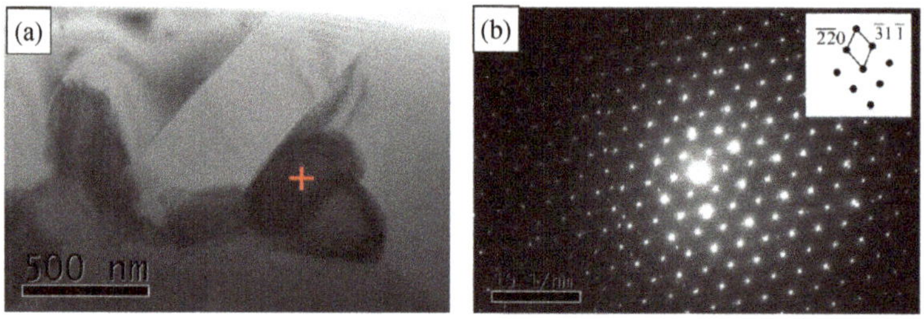

Figure 4. TEM image of precipitates of M_6C (**a**) and corresponding diffraction pattern (**b**) in the sample aged at 850 °C for 15 h.

3.3. IGC Behavior

DL-EPR test was employed to evaluate the degrees of sensitization of the samples aging treated at 850 °C. The representative DL-EPR curves are shown in Figure 5. The current density peaks in the anodic scan loop (I_a) and reverse scan loop (I_r) were determined by the curves. Then, the ratio of I_r/I_a can be obtained to evaluate the sensitization of IGC. Table 2 lists the determined values for the parameters I_a, I_r, and I_r/I_a. The I_a values varied little with the aging time. However, the I_r increased along with the aging. Since a new solution was used, the degree of sensitization was used to compare the relative IGC resistance of the samples. It was found that the degree of sensitization of the tested C276 superalloy increases with the aging time. This corresponds to the precipitate observation results, in which the number of precipitates increases with the progression of the aging treatment. Here, a new solution was used to evaluate the IGC behavior of C276 superalloy, which resulted in a large difference between these results and those reported previously [20]. The values in Table 2 indicate that this solution can reflect the degree of sensitization.

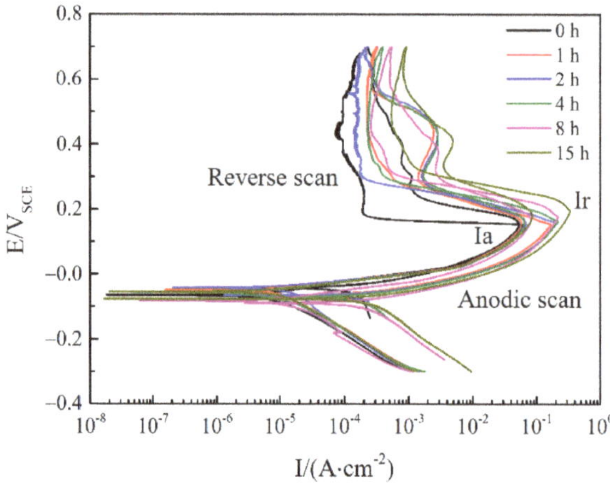

Figure 5. DL-EPR curves of C276 superalloy after aging treated at 850 °C for different times.

Table 2. Summary of I_a/I_r, and I_r/I_a determined based on the DL-EPR tests.

Aging Time	I_a (A·cm^{-2})	I_r (A·cm^{-2})	I_r/I_a
0 h	0.052304	0.065186	1.246
1 h	0.056988	0.16517	2.898
2 h	0.066707	0.1994	2.989
4 h	0.063476	0.21418	3.374
8 h	0.079404	0.27391	3.450
15 h	0.084394	0.33562	3.977

Figure 6 shows the microstructure of samples after DL-EPR tests. The as-received sample shows homogeneous etched morphology after testing. However, for the aged sample, severe corrosion at grain boundaries can be easily defined, as shown in Figure 6b. This indicates that C276 superalloy is sensitive to IGC attack after aging treatment, which agrees well with the DL-EPR curve analysis results.

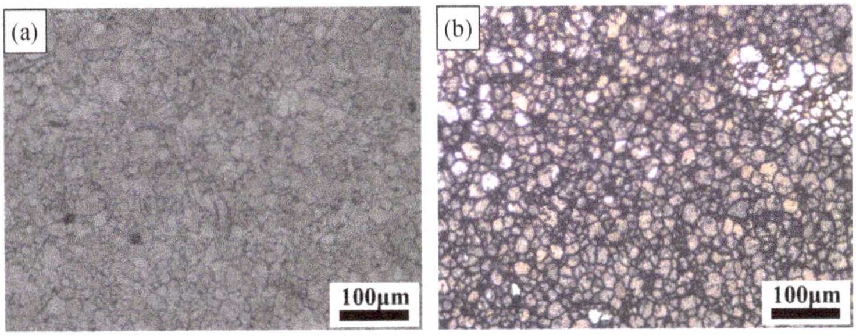

Figure 6. Optical microstructural observation of the samples after DL-EPR tests: (**a**) as-received material; (**b**) the sample aged at 800 °C for 12 h.

The precipitates of M_6C carbides after aging treatment are rich in Mo and W as detected by EDS analysis. This results in the formation of Mo- and W-depleted regions adjacent to the grain boundaries [29]. These regions are supposed to be easily attacked by corrosion during etching [30–32]. This implies that the precipitation of carbides during aging at high temperatures for higher than 4 h is detrimental to the IGC resistance of C276 superalloy. Moreover, the grain boundaries show different corrosion resistances in Figure 6b. Stratulat and Luo et al. have noticed this phenomenon in stainless steel and aluminum alloy [33,34]. They showed that the grain boundaries with high stored energy, which is related to the grain misorientation angle and dislocation density, were easily IGC attacked.

3.4. Microhardness and Wear Behavior

The microhardness of C276 superalloy after aging treatment at 850 °C was measured with the results displayed in Figure 7. The microhardness is shown to increase from about 330 to 360 HV after aging treatment for 4 h, and then declines very slightly with further increasing the aging time. The aging treatment at 800 °C led to very similar microhardness in the alloy with the aging treatment at 850 °C. Coupling the analysis with the precipitation observations in Figure 3, the precipitation of M_6C carbides is supposed to increase the microhardness of the aged C276 superalloy. Such a phenomenon of age hardening has also been found in a C276 alloy coatings by Mulligan and co-authors [35], who interpreted the increase in hardness to the formation of precipitates rich in Mo and Cr. It is worth noting that precipitation mainly occurs at grain boundaries and there is a slight grain growth during aging treatment, as evident from Figure 2; thus, a very small reduction in the microhardness is observed when the alloy is aging treated after 15 h.

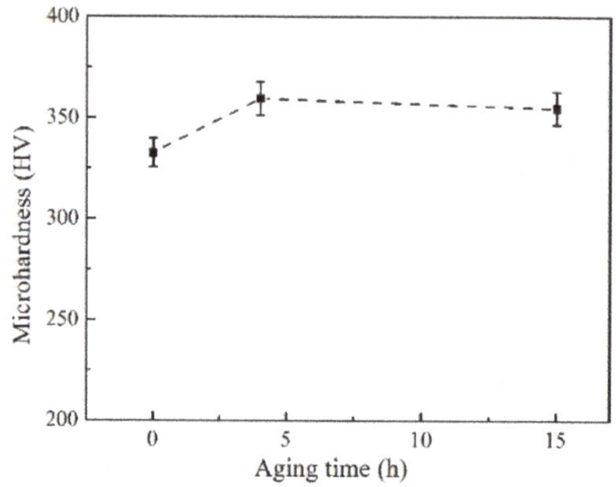

Figure 7. Microhardness of C276 superalloy after aging treatment at 850 °C with different aging times.

In previous studies [21–27,36–38], it has been reported that the wear behavior of a metallic material can be affected by many factors, such as its microstructure, phase constitution, mechanical properties, and the sliding wear conditions (including normal load, velocity, and atmosphere). Based on the precipitation results in Figure 2 and microhardness results in Figure 7, the C276 superalloy samples after aging treatment at 850 °C for 4 and 15 h were specifically selected for further wear property evaluations. Figure 8a shows the typical friction coefficient evolution histories during sliding wear as a function of time. The results reveal that the friction coefficient increases significantly during the early stage of sliding wear and then changes very slightly, indicating a relatively steady-state wear process. Figure 8b shows the average friction coefficient as a function of microhardness. Of note, the average friction coefficient in this paper is calculated by averaging the results obtained during the steady state wear stage from all three repeated wear tests for each condition. It can be seen from the figure that the average friction coefficient of C276 superalloy without aging treatment is about 0.6. The value is consistent with the typical friction coefficients for dry sliding wear of metallic materials [37]. It is interesting to observe that the friction coefficient increases first after aging treatment for 4 h and then drops with further increasing the aging time. The friction coefficient evolution tendency is consistent with that of microhardness. As shown in Figure 8b, the average friction coefficient of C276 superalloy increases when its microhardness is increased. The maximum friction coefficient of about 0.95 is achieved in the sample that has the greatest microhardness of about 360 HV corresponding to the aging treatment at 850 °C for 4 h. These increased friction coefficients in Figure 8 should be attributed to the formation of hard precipitates as displayed in Figure 2. It is obvious that the presence of precipitates not only results in the increase in friction coefficient, but also leads to the fluctuations of friction coefficient during steady state wear process in Figure 8a.

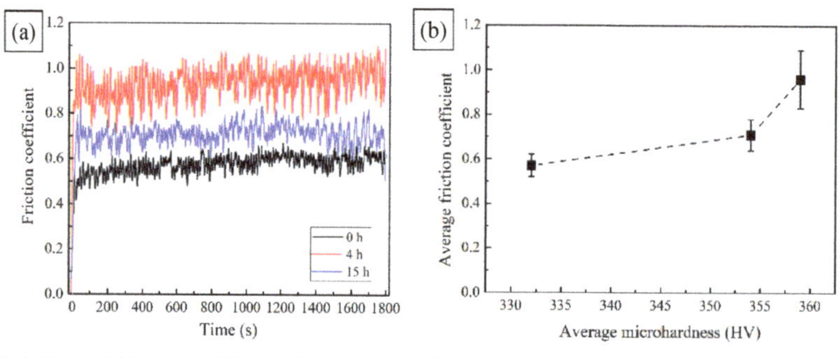

Figure 8. (**a**) Variations of friction coefficient of the samples during wear test, and (**b**) the relation between the friction coefficient and microhardness.

In order to reveal the influence of aging treatment on the wear resistance and wear mechanisms of C276 superalloy, the 3D wear track surface morphologies after wear test are characterized with the results shown in Figure 9. Obvious furrow marks associated with wear debris can be found in all the samples as labelled in the figures, indicating the severe abrasion wear. The samples with and without aging treatment show a great difference in the morphologies of furrow. Narrow and shallow wear marks develop in the C276 superalloy without aging treatment. On the contrary, remarkably wider and deeper grooves are found in the aged treated samples. To further analyze the surface morphologies after wear, the surface profiles were detected and drawn together in Figure 10. The wear track width and depth are about 373 μm and 5.5 μm, respectively, for the as-received alloy while increasing to about 1490–1530 μm and 21–22 μm, respectively, for the aged samples. It is evident that aging treatment of C276 superalloy leads to much larger wear tracks, indicating the obvious reductions of its wear resistance.

Although the microhardness increases after aging treatment, the wear resistance was reduced. Based on the observations on the wear tracks in Figure 9, it is estimated that the M_6C carbides may fall off from the material and increase the wear volume in the form of a third body during abrasive friction. This effect is different from the fine precipitates in the matrix, in which the fine scale precipitates can improve the wear resistance in the form of precipitation strengthening [38–40]. The increase in microhardness of C276 superalloy after aging treatment is caused by the formation of M_6C carbides. However, these precipitates mainly formed at grain boundaries, as confirmed in Figures 2–4. The precipitation at grain boundaries could deteriorate the mechanical properties of the matrix. Therefore, to preserve the good wear resistance of the C276 superalloy, the selection of aging treatment temperature and time should be very carefully selected. It is worth noting that this study is the first report considering the influence of aging treatment on the wear behavior of C276 superalloy, and a more comprehensive investigation on the mechanisms and sliding-wear-induced microstructure changes will be reported in our future research.

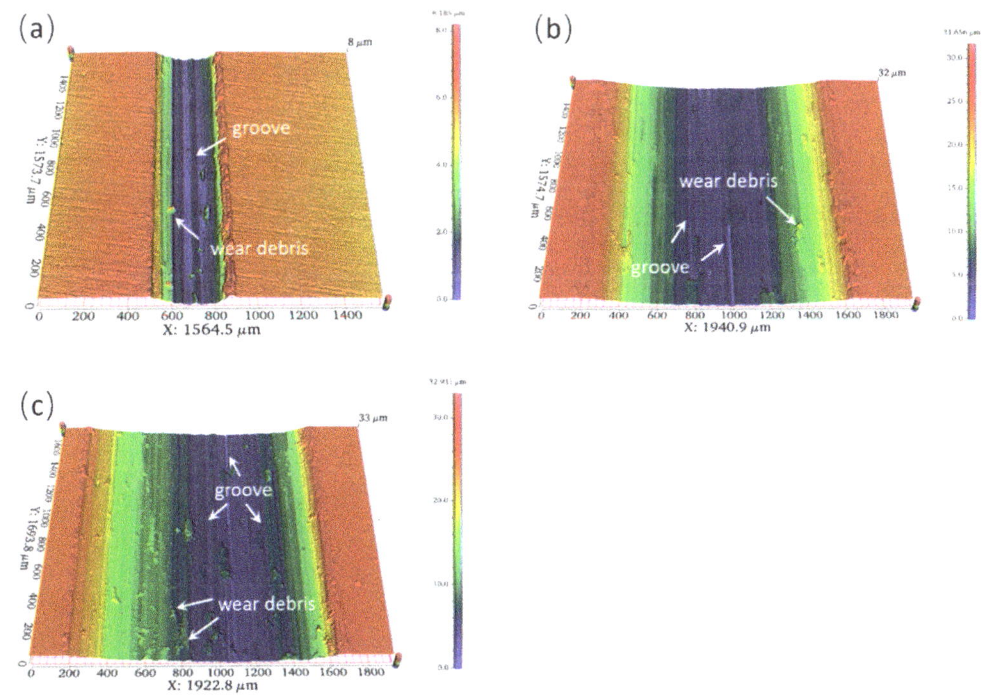

Figure 9. Topology profiles of the samples after wear test: (**a**) as-received sample; (**b**) the sample aged at 850 °C for 4 h; (**c**) the sample aged at 850 °C for 15 h.

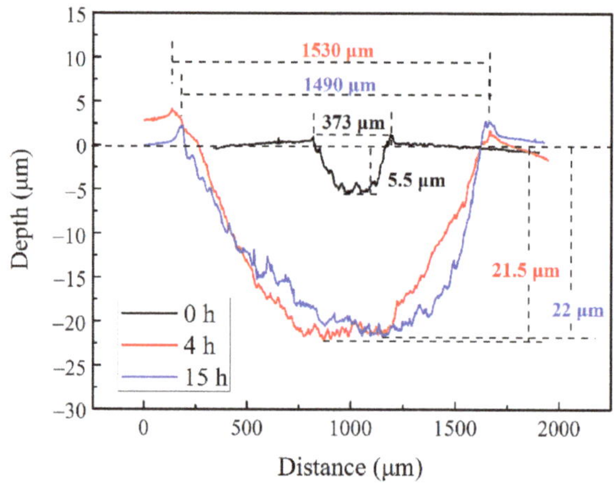

Figure 10. Comparisons of the wear track cross-sectional dimension of C276 superalloy after aging treatment at 850 °C with different aging time.

Based on the above analysis, it is found that the precipitation, corrosion resistance, microhardness, friction coefficient, and wear resistance of C276 superalloy can be greatly affected by aging treatment. Particularly, both the corrosion resistance and wear resistance deteriorated obviously for the C276 superalloy after aging longer than 4 h. Since this superalloy is mainly used in nuclear industry which faces cyclic load and erosion of coolant, long service life is important. Therefore, to avoid the failure of C276 superalloy components, a solution treatment is desired when the processing is conducted at high temperatures for a long time.

4. Conclusions

1. The equilibrium phase diagram for a Ni-Cr-Mo-based C276 superalloy was investigated using thermodynamic software Thermal-Calc. Coupled analysis with the precipitate observation, the M_6C carbides (which are rich in Mo and Cr) were found to precipitate at grain boundaries after aging treatment at 800–850 °C for higher than 4 h. μ phase and P phase had not been detected.
2. DL-EPR tests were used to estimate the IGC resistance of the aging-treated C276 superalloy. It is found that the degree of sensitization increased with the aging treatment time. It is supposed that the precipitation causes the formation of a Cr, W depleted zone adjacent to grain boundaries, deteriorating the corrosion resistance.
3. The microhardness of C276 superalloy increases after aging treatment, which was mainly attributed to the formation of hard M_6C precipitates. With further increasing the aging treatment time, a slight decrease in the microhardness was observed.
4. For the first time, sliding wear results indicated that both the friction coefficient and wear track dimension increased when C276 superalloy was aging treated. The hard precipitates fell off and acted as the third body during wear process, leading to a great reduction in the abrasion wear resistance.

Author Contributions: Conceptualization, C.Z. and Z.H.; methodology and experiments, P.Z. and X.Z.; writing—original draft preparation, P.Z. and X.Z.; writing—review and editing, C.Z. and Z.H. All authors have read and agreed to the published version of the manuscript.

Funding: This research was funded by the National Key Research and Development Program of China (No. 2019YFA0705300), the Joint Research Fund of Natural Science Foundation of Liaoning-the State Key Laboratory of Rolling and Automation, Northeastern University (2019KF0506) and the Fundamental Research Funds for the Central Universities of China (DUT19JC26).

Data Availability Statement: Not available.

Conflicts of Interest: The authors declare no conflict of interest.

References

1. Li, Z.; Han, J.S.; Lu, J.J.; Chen, J.M. Cavitation erosion behavior of Hastelloy C-276 nickel-based alloy. *J. Alloys Compd.* **2015**, *619*, 754–759. [CrossRef]
2. Zhu, Y.; Hou, J.; Yu, G.; Qiu, J.; Chen, S.; Zhou, X. Effects of exposing temperature on corrosion performance of weld joint of a Ni–Mo–Cr alloy. *J. Fluor. Chem.* **2016**, *182*, 69–75. [CrossRef]
3. Liu, M.; Zheng, J.; Lu, Y.; Li, Z.; Zou, Y.; Yu, X.; Zhou, X. Investigation on corrosion behavior of Ni-based alloys in molten fluoride salt using synchrotron radiation techniques. *J. Nucl. Mater.* **2013**, *440*, 124–128. [CrossRef]
4. Wanderka, N.; Bakai, A.; Abromeit, C.; Isheim, D.; Seidman, D.N. Effects of 10MeV electron irradiation at high temperature of a Ni–Mo-based Hastelloy. *Ultramicroscopy* **2007**, *107*, 786–790. [CrossRef] [PubMed]
5. Raghavan, M.; Berkowitz, B.J.; Scanlon, J.C. Electron microscopic analysis of heterogeneous precipitates in Hastelloy C-276. *Metall. Trans. A* **1982**, *13 A*, 979–984. [CrossRef]
6. Ahmad, M.; Akhter, J.I.; Akhtar, M.; Iqbal, M.; Ahmed, E.; Choudhry, M.A. Microstructure and hardness studies of the electron beam welded zone of Hastelloy C-276. *J. Alloys Compd.* **2005**, *390*, 88–93. [CrossRef]
7. Ernesto, V.A.R.T.; Guastaldi, A.C. Investigation of the metal/porcelain interface in LASER-welded Ni-Cr-Mo alloy. *Weld. Int.* **2009**, *23*, 193–199. [CrossRef]
8. Guo, Y.Q.; Wu, D.J.; Ma, G.Y.; Guo, D.M. Trailing heat sink effects on residual stress and distortion of pulsed laser welded Hastelloy C-276 thin sheets. *J. Mater. Process. Technol.* **2014**, *214*, 2891–2899. [CrossRef]
9. Zhang, C.; Zhang, L.; Shen, W.; Liu, C.; Xia, Y. The Kinetics of Metadynamic Recrystallization in a Ni-Cr-Mo-Based Superalloy Hastelloy C-276. *J. Mater. Eng. Perform.* **2016**, *25*, 545–552. [CrossRef]
10. Zhang, C.; Zhang, L.W.; Shen, W.F.; Li, M.F.; Gu, S.D. Characterization of Hot Deformation Behavior of Hastelloy C-276 Using Constitutive Equation and Processing Map. *J. Mater. Eng. Perform.* **2014**, *24*, 149–157. [CrossRef]
11. Zhu, Z.; Zhang, L.; Gu, S. Modeling effects of process parameters on vacuum hot bulge forming of rotor cans. *Rare Metal Mater. Eng.* **2014**, *43*, 2257–2261.
12. Zhu, Z.; Zhang, L.; Gu, S. FEM analysis of vacuum hot bulge forming of Hastelloy C-276 thin-walled cylindrical workpiece. *J. Cent. South Univ.* **2014**, *21*, 3019–3023. [CrossRef]
13. Zhan, L.; Lin, J.; Dean, T.A. A review of the development of creep age forming: Experimentation, modelling and applications. *Int. J. Mach. Tools Manuf.* **2011**, *51*, 1–17. [CrossRef]
14. Unocic, K.A.; Shin, D.; Sang, X.; Cakmak, E.; Tortorelli, P.F. Single-step aging treatment for a precipitation-strengthened Ni-based alloy and its influence on high-temperature mechanical behavior. *Scr. Mater.* **2019**, *162*, 416–420. [CrossRef]
15. Ni, M.; Liu, S.; Chen, C.; Li, R.; Zhang, X.; Zhou, K. Effect of heat treatment on the microstructural evolution of a precipitation-hardened superalloy produced by selective laser melting. *Mater. Sci. Eng. A* **2019**, *748*, 275–285. [CrossRef]
16. Liu, J.; Zhang, J.; Lu, Y.; Li, X.; Li, Z.; Zhou, X. Effect of long-term aging on microstructure and mechanical properties of alloy C276. *Acta Metall. Sin.* **2013**, *49*, 763–768. [CrossRef]
17. Akhter, J.I.; Shaikh, M.A.; Ahmad, M.; Iqbal, M.; Shoaib, K.A.; Ahmad, W. Effect of aging on the hardness and impact properties of Hastelloy C-276. *J. Mater. Sci. Lett.* **2001**, *20*, 333–335. [CrossRef]
18. Yang, J.; He, Y.; Qin, C.; Zhao, W.; Chen, S.; Gao, Z. Microstructure evolution in a Ni–Mo–Cr superalloy subjected to simulated heat-affected zone thermal cycle with high peak temperature. *Mater. Des.* **2015**, *86*, 230–236. [CrossRef]
19. He, Y.M.; Yang, J.G.; Chen, S.J.; Li, Z.; Gao, Z.L. Effect of high-temperature aging on microstructure and mechanical properties of Ni–Mo–Cr based superalloy subjected to simulated heat-affected zone thermal cycle. *J. Alloys Compd.* **2016**, *660*, 266–275. [CrossRef]
20. Zhang, C.; Zhang, L.; Cui, Y.; Feng, Q.; Cheng, C. Effects of High-Temperature Aging on Precipitation and Corrosion Behavior of a Ni-Cr-Mo-Based Hastelloy C276 Superalloy. *J. Mater. Eng. Perform.* **2020**, *29*, 2026–2034. [CrossRef]
21. Chen, J.; Wang, J.; Yan, F.; Zhang, Q.; Li, Q.A. Corrosion wear synergistic behavior of Hastelloy C276 alloy in artificial seawater. *Trans. Nonferrous Met. Soc. China* **2015**, *25*, 661–668. [CrossRef]
22. Chen, J.; Wang, J.; Yan, F.; Zhang, Q.; Li, Q.A. Assessing the corrosion–wear behaviours of Hastelloy C276 alloy in seawater. *Lubr. Sci.* **2016**, *28*, 67–80. [CrossRef]
23. Wang, J.; Chen, J.; Chen, B.; Yan, F.; Xue, Q. Wear behaviors and wear mechanisms of several alloys under simulated deep-sea environment covering seawater hydrostatic pressure. *Tribol. Int.* **2012**, *56*, 38–46. [CrossRef]
24. Hashim, M.; Duraiselvam, M. Enhancing Tribological and Corrosion Resistance of Hastelloy C-276 through Laser Surface Treatment. *Mater. Sci. Forum* **2015**, *830–831*, 659–662.
25. Yilbas, B.S.; Ali, H. Laser texturing of Hastelloy C276 alloy surface for improved hydrophobicity and friction coefficient. *Opt. Lasers Eng.* **2016**, *78*, 140–147. [CrossRef]
26. Li, K.M.; Song, K.J.; Guan, J.; Yang, F.; Liu, J. Tribocorrosion behavior of a Ti6Al4V alloy electromagnetic induction nitride layer in a fluorine-containing solution. *Surf. Coat. Technol.* **2020**, *386*, 125506.
27. Deng, G.; Zhao, X.; Su, L.; Wei, P.; Zhang, L.; Zhan, L.; Chong, Y.; Zhu, H.; Tsuji, N. Effect of high pressure torsion process on the microhardness, microstructure and tribological property of Ti6Al4V alloy. *J. Mater. Sci. Technol.* **2021**, *94*, 183–195.

28. Sun, W.; Qin, X.; Guo, J.; Lou, L.; Zhou, L. Microstructure stability and mechanical properties of a new low cost hot-corrosion resistant Ni–Fe–Cr based superalloy during long-term thermal exposure. *Mater. Des.* **2015**, *69*, 70–80.
29. Jiao, S.; Zhu, G.; Dong, J.; Zhang, Q. Carbide evolution and Mo depletion law in Hastelloy C-276. *J. Mater. Eng.* **2011**, *1*, 47–52.
30. Krupp, U. Improving the resistance to intergranular cracking and corrosion at elevated temperatures by grain-boundary-engineering-type processing. *J. Mater. Sci.* **2008**, *43*, 3908–3916.
31. Lee, H.T.; Wu, J.L. Intergranular corrosion resistance of nickel-based alloy 690 weldments. *Corros. Sci.* **2010**, *52*, 1545–1550. [CrossRef]
32. Liu, X.Y.; Li, M.J.; Gao, F.; Liang, S.X.; Zhang, X.L.; Cui, H.X. Effects of aging treatment on the intergranular corrosion behavior of Al-Cu-Mg-Ag alloy. *J. Alloys Compd.* **2015**, *639*, 263–267. [CrossRef]
33. Stratulat, A.; Duff, J.A.; Marrow, T.J. Grain boundary structure and intergranular stress corrosion crack initiation in high temperature water of a thermally sensitised austenitic stainless steel, observed in situ. *Corros. Sci.* **2014**, *85*, 428–435. [CrossRef]
34. Luo, G.; Xiao, H.; Li, S.; Wang, C.; Zhu, Q.; Song, L. Quasi-continuous-wave laser surface melting of aluminium alloy: Precipitate morphology, solute segregation and corrosion resistance. *Corros. Sci.* **2019**, *152*, 109–119. [CrossRef]
35. Mulligan, C.P.; Wei, R.; Yang, G.; Zheng, P.; Deng, R.; Gall, D. Microstructure and age hardening of C276 alloy coatings. *Surf. Coat. Technol.* **2015**, *270*, 299–304. [CrossRef]
36. Deng, G.; Tieu, A.K.; Su, L.; Wang, P.; Wang, L.; Lan, X.; Cui, S.; Zhu, H. Investigation into reciprocating dry sliding friction and wear properties of bulk CoCrFeNiMo high entropy alloys fabricated by spark plasma sintering and subsequent cold rolling processes: Role of Mo element concentration. *Wear* **2020**, *460–461*, 203440. [CrossRef]
37. Deng, G.; Tieu, A.K.; Lan, X.; Su, L.; Wang, L.; Zhu, Q.; Zhu, H. Effects of normal load and velocity on the dry sliding tribological behaviour of CoCrFeNiMo0.2 high entropy alloy. *Tribol. Int.* **2020**, *144*, 106116. [CrossRef]
38. Wang, L.; Tieu, A.K.; Cui, S.; Deng, G.; Wang, P.; Zhu, H.; Yang, J. Lubrication mechanism of sodium metasilicate at elevated temperatures through tribo-interface observation. *Tribol. Int.* **2020**, *142*, 105972. [CrossRef]
39. Zhou, Z.; Shan, Q.; Jiang, Y.; Li, Z.; Zhang, Z. Effect of nanoscale V_2C precipitates on the three-body abrasive wear behavior of high-Mn austenitic steel. *Wear* **2019**, *436–437*, 203009. [CrossRef]
40. Yang, K.; Yu, S.; Li, Y.; Li, C. Effect of carbonitride precipitates on the abrasive wear behaviour of hardfacing alloy. *Appl. Surf. Sci.* **2008**, *254*, 5023–5027. [CrossRef]

metals

MDPI

Article

Effect of Mo Element on the Mechanical Properties and Tribological Responses of CoCrFeNiMo*x* High-Entropy Alloys

Ying Liu [1,†], Yongxin Xie [1,†], Shaogang Cui [1,2,*], Yanliang Yi [1,2,*], Xuewei Xing [1], Xiaojian Wang [1] and Wei Li [1]

1 Institute of Advanced Wear & Corrosion Resistant and Functional Materials, Jinan University, Guangzhou 510632, China; liuying2000@jnu.edu.cn (Y.L.); xieyongxin@stu2018.jnu.edu.cn (Y.X.); xxwhym1228@stu2019.jnu.edu.cn (X.X.); xiaojian.wang@jnu.edu.cn (X.W.); Lwxasn@sohu.com (W.L.)

2 Shaoguan Research Institute of Jinan University, 168 Muxi Avenue, Shaoguan 512027, China

* Correspondence: sgcui@jnu.edu.cn (S.C.); ylyi192@jnu.edu.cn (Y.Y.); Tel.: +86-13802425607 (S.C.); +86-15091339192 (Y.Y.)

† Ying Liu and Yongxin Xie contributed equally to this work.

Abstract: Certain amounts of precipitate in CoCrFeNiMo*x* (simplified as Mo*x*) is beneficial to the wear resistance; however, the optimal chemical content of Mo and the anti-wear mechanism behind it remains unclear. The Mo*x* (x = 0, 0.3, 0.5, 1, 1.5 in molar ratio) high entropy alloys (HEAs) were manufactured, the evolution of their microstructure, mechanical, friction, and wear properties with Mo content was studied. The results displayed that the mechanical properties of the FCC solid solution were enhanced from Mo0 to Mo0.3, then kept unchanged till $x = 1.5$. The volume fraction of the precipitates increased with Mo content. The Mo1 presents the lower average friction coefficient and wear rate, attributed to the desired types, amount, size, distribution of the hard σ and μ phases in the ductile FCC solid solution. The detailed mechanism behind their tribological behaviors were discussed in the manuscript.

Keywords: high-entropy alloy; microstructure; mechanical properties; CoCrFeNiMo*x*; friction and wear

Citation: Liu, Y.; Xie, Y.; Cui, S.; Yi, Y.; Xing, X.; Wang, X.; Li, W. Effect of Mo Element on the Mechanical Properties and Tribological Responses of CoCrFeNiMo*x* High-Entropy Alloys. *Metals* **2021**, *11*, 486. https://doi.org/10.3390/met11030486

Academic Editor: Jiro Kitagawa

Received: 13 February 2021
Accepted: 5 March 2021
Published: 15 March 2021

1. Introduction

High entropy alloys (HEAs), namely the multicomponent alloys, have become a hotspot of material research since 2004 when proposed by Yeh et al. [1] and Cantor et al. [2]. Because they possess high thermal stability, superior physical and excellent mechanical properties, such as high hardness [3], high strength, and high ductility, especially at a cryogenic temperature [4,5], excellent wear resistance [6], and corrosion resistance [7], which make the HEAs the promising materials as components in structural and mechanical materials, particularly at cryogenic temperature. These special features have been on account of the high mixing entropy, sluggish diffusion, and severe lattice distortion. HEAs are composed of five or more principal elements with the concentration of each element being between 5 wt.% and 35 wt.%, they have a stable solid-solution state attributed to the high mixing entropy $\Delta Smix$, and the distribution of the five constituent elements is relatively random and uniform [8].

HEAs such as CoCrFeNi alloy systems, with a face-centered (FCC) structure, have attracted many attentions due to their exceptional ductility, fracture toughness, and high work hardening capability [9]. Gludovatz et al. had researched the mechanical properties of CrMnFeCoNi HEA, the results show that it has better toughness and yield strength than most pure metals and alloys, and has the comparable strength as structural ceramics, closing to the strength of some bulk metallic glasses [10]. However, the HEA matrix alone, especially the single-phase catalytic cracking structure, is not enough for engineering applications at room temperature. For example, the single-faced CoCrFeNi HEA with an FCC structure has a yield strength of only 155 MPa and a tensile strength of 472 MPa [11].

As a result, Mo element is usually added to the single-phase CoCrFeNi FCC solid solution. to obtain a combination of high strength and excellent ductility, for pursuit solid solution strengthening and precipitation hardening. This is because the Mo atom has a large atomic size for both solid solutions caused by the higher lattice distortion and precipitation hardening [12,13]. Mo can promote the precipitation of hard and brittle (Cr,Mo)-rich σ and (Mo,Cr)-rich μ phases in a FCC solid solution. It will increase the strength and hardness but reduce the ductility [14]. When the molar ratio of Mo is beyond 0.3, it shows the emergence of a dendritic structure and forms hard intermetallic phases with Co, Cr, Fe, and Ni elements [15]. Liu et al. [11] had used a thermal calculation computer program to evaluate the phase stability of a complex multi-component system and calculated the (CoCrFeNi)-Mo pseudo-binary phase diagram. It suggests that with the increasing Mo element the microstructures of the as-cast Mo0.2 and Mo0.3 alloys develop to a dendritic microstructure, and Mo starts to segregate in the inter-dendritic areas in the as-cast Mo0.3 alloy showing a mixed two-phase structure (FCC + σ). When the molar ratio is beyond 0.3, then three phases (FCC+ σ + μ) form. The precipitation of these hard topologically close-packed (TCP) phases (σ and μ) with a highly high hardness of 15 GPa has a substantial influence on the mechanical properties of FeCoCrNi based high entropy alloys, which strongly affect their tribological behavior as well [11,14]. Li et al. had studied the fracture performance and fracture mechanism of CoCrFeNiMo0.2 high entropy alloy, revealing that the alloy possesses a good damage tolerance at room temperature [16]. Besides, the study of fracture mechanism shows that the boundary between the grain and intermetallic precipitation is the weak part of the microporous nucleation, and the cracks mainly propagate along the interface between grain boundary and precipitation particles. Niu et al. [17] investigated the as-annealed FeCoCrNiMo*x* high entropy alloys, revealing annealing treatment induces Mo addition to be unstable, and the phase separation happens. Bae et al. studied cold rolling and subsequent annealing treatment at different temperature of Co17.5Cr12.5Fe55Ni8Mo5 alloys, the structure is composed of FCC and μ phase, μ phase is the Mo-rich phase [18], but the mechanical/wear properties is less probing. Deng et al. had investigated the fraction of Mo on the wear resistance of FeCoCrNiMo*x* (x = 0~0.3) HEAs made by spark plasma sintering and cold rolling [19]. It had found that the wear resistance improves continuously with the increase of Mo content in the range of x = 0~0.3. The revealing of microstructure evolution in subsurface and near surface suggested that though some partially tribo-oxidized wear debris formed the loosely agglomerated patches, the dominant FCC solid solution failed to be tribo-oxidized. The direct metal/metal contact takes place. However, this study did not further explore the Mo molar fraction beyond 0.3, in which the hard-intermetallic precipitation forms. The excellent deformation ability makes the CoCrFeNi be an ideal FCC solid solution for the strengthening of various intermetallic compounds. The homogeneous precipitation of hard intermetallic particles is a powerful method to strengthen the steel and alloys due to their resistance to gliding dislocations, and the precipitation strengthening depends on the type, size, number density, and distribution of the precipitated phase. Therefore, the tribological behaviors of Mox with Mo molar fraction beyond 0.3 shall be studied to reveal the ideal elemental composition.

Mo is the σ/μ phase accelerating element. In order to obtain the FCC + σ + μ phase structure, molybdenum is added as a variable element, trying to explore the influence of the variation of σ/μ phase on the alloy structure and properties. This paper was designed to investigate the effect of Mo element on the tribological behavior of Mo*x* HEAs, where x was in a much wider range of 0~1.5). It aims to find the optimal Mo content at which the lower average friction coefficient and wear rate were simultaneously achieved. The effect of Mo content on the microstructure, mechanical properties and friction and wear properties of the HEAs alloys were also investigated in detail, and the anti-wear mechanism behind the optimal chemical compositions was revealed.

2. Materials and Methods

The CoCrFeNiMox (x = 0, 0.3, 0.5, 1, 1.5 in atomic proportion) alloys were prepared by vacuum arc melting method melted in a water-cooled copper mold under high purity argon atmosphere. For convenience, these prepared alloys are referred to as Mo0, Mo0.3, Mo0.5, Mo1, and Mo1.5, respectively.

High purity elements (≥99.9 wt.%) were used as raw materials. The FeCoCrNiMox ingots were remelted at least five times to improve the chemical homogenous distribution. The microstructures, physical and mechanical properties of these as-cast samples were characterized to evaluate the effect of Mo content on the microstructure and properties.

The friction and wear properties of the prepared Mox HEAs were evaluated using a reciprocating tribometer (Rtec Co. Ltd., San Jose, CA, USA) with a configuration of ball-on-plate. The schematic diagram of reciprocating ball-on-disc was shown in Figure 1. Before the testing, the test surfaces of fabricated samples were ground by 200#, 600#, 1000#, 2000# SiC paper and polished with 3 μm suspension, the surfaces have a surface roughness (Sa) of 0.034 μm, respectively. Then they were ultrasonically cleaned in alcohol. The YG 6 ceramic ball with a diameter of 5 mm and chemical composition of 94%WC and 6% Co was used as the counter sliding surface. The test conditions were set as the applied load of 20 N, duration of 40 min, a sliding stroke of 10 mm, reciprocating frequency of 5 HZ, which slides a total distance of 240 m. To ensure repeatability, each test was repeated at least three times. After tribological tests, the 3D surface profiles of the wear tracks on plates were measured by the 3D profiler (Rtec Co. Ltd., San Jose, CA, USA).

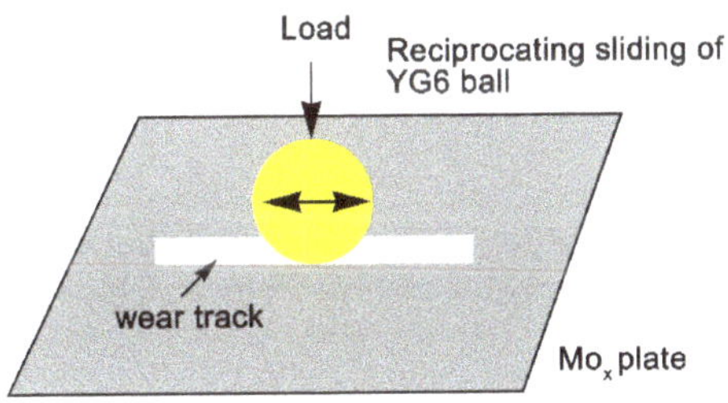

Figure 1. Schematic diagram of reciprocating ball-on-disc.

The structural features of HEAs were characterized by X-ray diffraction (XRD, Ultima IV, Rigaku, Tokyo, Japan) with Kα1 radiation at 40 kV/40 mA, scanning from 30 to 60° in 2θ at a scanning speed of 5°/min. The microstructure was investigated by scanning electron microscope (SEM, Phenom XL, The Netherlands) operated at 15 kV with an energy dispersive spectrometer (EDS), and the Image Pro Plus software was used to measure the fraction of the precipitate phases.

The hardness was measured on the polished specimens with a microhardness tester (HXD-1000TMSC/LCD, Shanghai, China) at a load of 1.961 N and dwell time of 15 s. To assure accuracy, each alloy was measured with at least 5 points for averaging.

The nanoindentation hardness tests were performed on the Nanoindenter (Nano Indenter G200, Keysight Inc., Santa Rosa, CA, USA) with a Berkovich indenter at room temperature, with a loading time of 15 s, hold time of 2 s, Poisson's ratio of 0.28. The allowable drift rate was set to 0.1 nm/s, and the surface approach velocity to 10 nm/s. NanoSuite software (version 6.2, Santa Rosa, CA, USA) was used to set these parameters and collect the results. The peak load of 10 mN is used to measure the hardness of the different phases in the Mox HEA. An array of 10 × 10 indentation points was carried out for each sample. Then SEM was used to image these points, from which qualified points were selected.

The compression tests were performed using a cylinder specimen with the size of Ø 5 mm × 10 mm on an MTS E45.305 electronic universal material testing machine (MTS, Shanghai, China) at a compression rate of 1 mm/min at room temperature. The surface of the test samples was polished down to a 2000-grit SiC paper to eliminate scratches. Each compression test was repeated three times to ensure repeatability.

3. Results

3.1. Microstructure

Figure 2 is the XRD patterns of the as-cast Mox HEAs. It can be seen that all samples have a basic FCC solid solution structure. It can be seen that at $x = 0$, the FeCoCrNi alloy have a single FCC structure. Some XRD peaks belong to the σ phase appears when $x \geq 0.3$. When $x \geq 1$, Mox alloys exhibits an FCC + σ + μ multiphase structure. Figure 3 presents the corresponding SEM backscattering micrographs of the microstructure of these Mox HEAs and the variation of volume fraction of the precipitates with the Mo molar ratio was given in Figure 3f. Table 1 presents the chemical compositions of the different phases in Mox measured by EDS, it revealed that a significant amount of Mo and Cr was in σ phase due to their high molar ratio, especially for Mo1 and Mo1.5, the Mo: Fe molar ratio reaches 1.64 and 2.28, respectively, and even higher molar ratio was in the bright μ phase. Figure 4 presents the EDS results of the different phases (labeled as "*") in Mox. For Mo0, because the Fe (1.27 Å), Co (1.25 Å), Cr (1.27 Å), and Ni (1.24 Å) atoms have similar atomic sizes, and there is no significant negative or positive atom pair mixing enthalpy between each element atoms [20], therefore, it has a single FCC solid solution structure. The bulky grains have an average diameter of 53 $\pm$ 15 μm. When the Mo molar ratio grows to 0.3, the diffraction peak beside the FCC (111) was detected, in combination with the SEM image in Figure 3b EDS analysis in Figure 4, this peak is belonging to the σ phase. Table 2 displayed the crystallite sizes and lattice parameters of σ/μ phases. The σ phase is indexed as the tetragonal structure (a = 9.2443 Å, c = 4.7921 Å, c/a = 0.5183) and the σ phase is indexed as the hexagonal structure (a = 3.5778 Å, c = 25.7538 Å, c/a = 7.1982). Further detailed analysis in Figure 3b reveals that the fraction of the precipitates was estimated to be about 3.2%. Such σ phase is usually hard and brittle, the existence of σ phase will deteriorate the inter-grain embrittlement, which is normally deleterious to the mechanical properties [21], but it could be beneficial to the abrasion resistance to some extent due to the superior hardness [11]. When the Mo molar ratio continues to be increased, the secondary phase precipitates more evidently in the Mo0.5 alloy, which includes the σ and μ phases, as indicated in the inset of Figure 3c. The total volume fraction of the secondary phases was estimated to be 14.8%, including the 12.1% σ and 2.7% μ phases. It was also noted that the μ phase was formed on the fringes of the σ phase, which was reported that it was formed through the transformation the σ phase during the solidification [14]. Figure 4 reveals that the μ phase is rich in Mo and some Cr, and the σ phase in Cr and some Mo element. It is obvious that the size and number of the secondary phase increase significantly for Mo1 and Mo1.5, displaying more dendritic structure, 48% secondary phases including 25% σ and 23% μ was estimated for Mo1. With the Mo molar ratio being increased to 1.5, and the secondary phase turns out to be μ phase dominantly with the diameter between 29–81 μm and the fraction of the secondary phase is estimated to be 77.5%. Mo1.5 alloy exhibits more μ phase and less σ phase compared with Mo1 alloy, which is because the excess molybdenum content makes for converting the σ phase into μ phase [14].

The lattice distortion was also introduced by the Mo addition. As shown in Figure 2, the main diffraction peak for FCC (111) plane locates at 43.5° for $x = 0$. It deviates to 43.1° for $x = 0.3$, this is because the atomic radius of Mo (1.39 Å) is larger, compared with the other four constitutional elements, the addition of Mo into the FCC solid solution will cause pronounced lattice distortion. When the Mo molar ratio continues to increase, the diffraction peak shows little change. This indicates that the content of Mo in the FCC solid solution lattice reaches the saturation state.

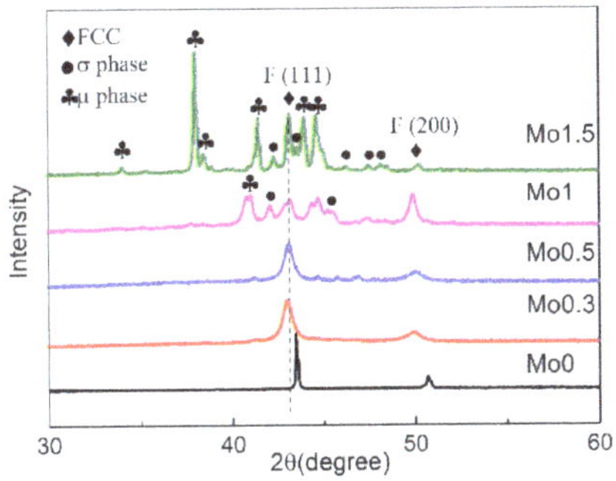

Figure 2. XRD patterns of the as-cast Mo*x* high entropy alloys.

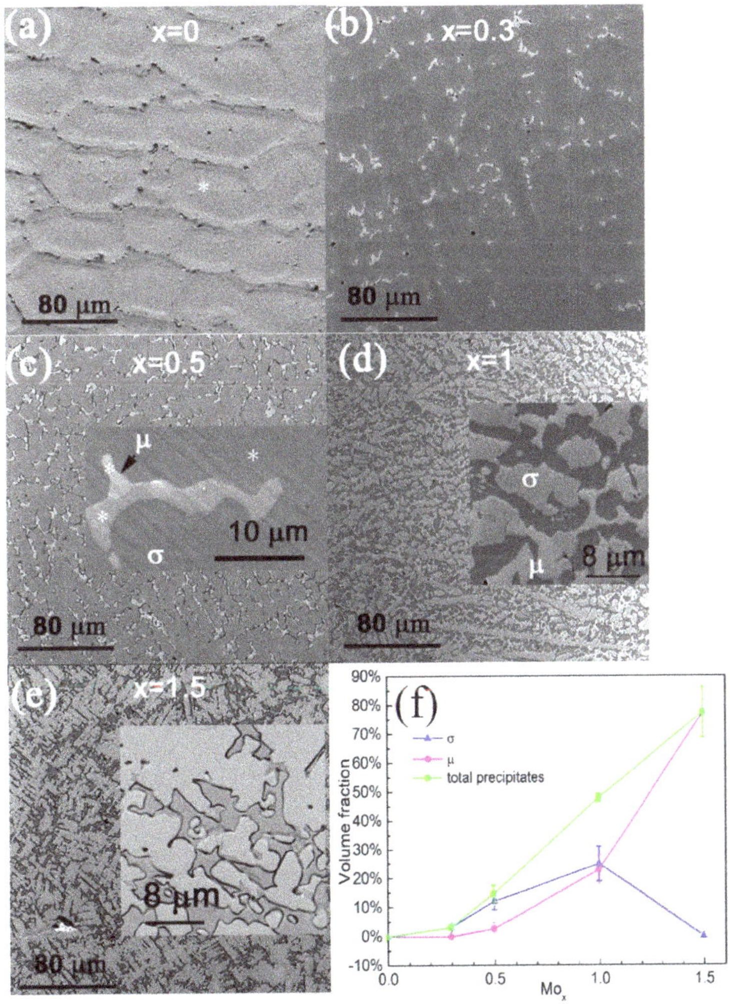

Figure 3. SEM backscattering images of the as-cast Mox high entropy alloys: (**a**) For Mo0, (**b**) for Mo0.3, (**c**) for Mo0.5, (**d**) for Mo1, (**e**) for Mo1.5, and (**f**) the variation of the volume fraction of the precipitates with the Mo molar ratio.

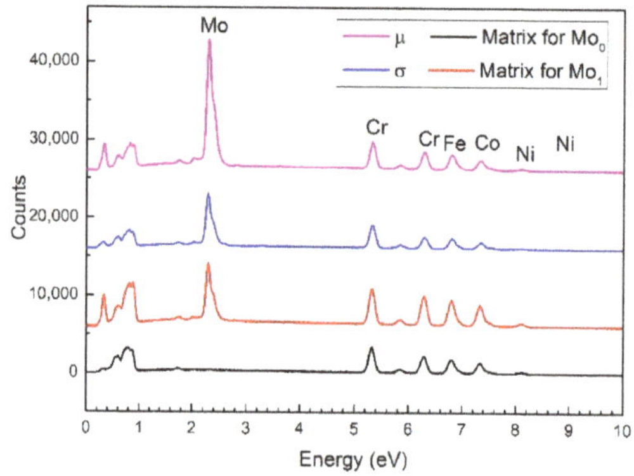

Figure 4. EDS analysis of the elements for different phases.

Table 1. Chemical composition of the different phases in the studied compounds (normalized molar ratio).

HEAs	Analyzed Area	Normalized Atomic Molar Ratio by Dividing the Fe Atoms				
		Fe	Co	Cr	Ni	Mo
Mo0	FCC solid solution	1	1.01	0.99	1.04	0
Mo0.3	FCC solid solution	1	0.99	0.88	1	0.31
	σ	1	1.08	1.49	0.83	1.42
Mo0.5	FCC solid solution	1	0.98	0.84	1.09	0.47
	σ	1	1.04	1.34	0.89	1.59
Mo1	FCC solid solution	1	1.05	0.85	1.16	0.8
	σ	1	1.07	1.16	0.93	1.64
	μ	1	1.12	1.12	0.95	1.94
Mo1.5	FCC solid solution	1	0.96	0.79	1.19	0.96
	σ	1	1.05	1.01	0.96	2.28
	μ	1	1.0	1.12	0.88	2.56

Table 2. The crystallite sizes and lattice parameters of σ/μ phases.

Phase	Lattice Parameters			Crystallite Size
σ phase	a = 9.165 Å	c = 4.739 Å	c/a = 0.517 Å	16–33 μm
μ phase	a = 4.757 Å	c = 25.589 Å	c/a = 5.379 Å	29–81 μm

3.2. Nanoindentation Results

Figure 5a presents the load-depth curves of the different phases in the HEA and SEM images of these indents, please note that these indents were only some examples. The calculated results were listed in Table 3. Firstly, 10 × 10 indents were conducted, then they were imaged by SEM, from which qualified indents were selected. For example, representative indents for σ, μ, and FCC solid solution were shown in Figure 5b. For the FCC solid solution, it can be found that the Mo0.3–1.5 HEA has a similar load-depth curve, indicating similar mechanical properties. Mo0 has a different load-depth curve, which shows a larger depth than Mo0.3–1.5. This represents the Mo0 FCC solid solution has a weak penetration resistance.

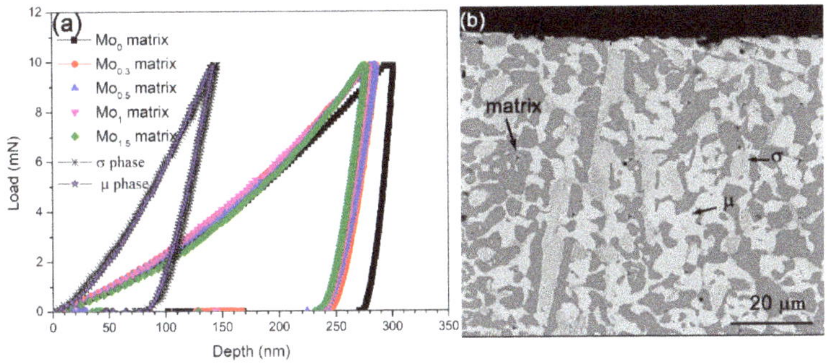

Figure 5. (**a**) Load-depth curves for the FCC solid solution, σ and μ phases in the Mo*x* HEA for the maximum load of 10 mN and (**b**) a given SEM backscattering image of these indents for Mo1.

Table 3. The nano-hardness of the prepared materials in different phases.

Materials	Hardness (GPa)
YG6 ball	20 ± 0.5
Mo0 FCC solid solution	3.8 ± 0.1
FCC solid solution for Mo0.3, Mo0.5, and Mo1.5	4.7 ± 0.2
σ phase	21 ± 1.2
μ phase	21 ± 1.7

Table 3 shows the nanohardness of the prepared Mo*x* HEAs measured by nanoindenter. A heavy load of 300 mN was applied to measure the global mechanical response of YG 6. It found that the counter YG6 ball has a hardness of 20 ± 0.5 GPa. Then a light load of 10 mN was used to measure the mechanical response of the separate phases in the Mox HEA. From Figure 5 and Table 3, it can be found that the μ has a similar hardness to that of the σ phase. They were also similar to that of YG 6 ball in hardness. The Mo-saturated FCC solid solution for Mo0.3–1.5 has a hardness 0.9 GPa higher than that of Mo0 FCC solid solution. The nanoindentation results prove that the mechanical strength of the matrix strengthened by solid solution reaches the maximum when the molar ratio of Mo increases to 0.3. Further increase in Mo molar ratio contributes little to the increase of mechanical strength of the matrix, but it will increase volume fraction of the precipitated intermetallic. This becomes the dominant factor in determining mechanical properties.

3.3. Mechanical Properties of Fabricated Mox

Figure 6a presents the compressive engineering stress–strain relationship and the microhardness of the HEAs. A high load of 1.961 N was adopted to measure the microhardness, therefore, it reflects the global mechanical response of the Mo*x* HEA. The Mo0 presents low compressive yield stress of 207 MPa and hardness of 161.4 HV but shows excellent ductility. When the Mo molar ratio increases to 0.5, with the intermetallic precipitated increasingly, the yield stress and hardness increase steadily, and the ductility does not show pronounced deterioration. When the Mo molar ratio increases to 1, the yield stress and hardness increase to 825 MPa and 550.2 HV, respectively, but the ductility also shows an obvious decline. The fracture stress is 1822.4 MPa, in a strain of 11.4%. The Mo1.5 presents an obvious brittle fracture nature with the fracture stress of 996 MPa and hardness of 779 HV, not presenting any plastic deformation behavior. Such superior hardness and inferior ductility might be due to the overly high fraction volume of the precipitated intermetallic, as shown in Figure 3e,f.

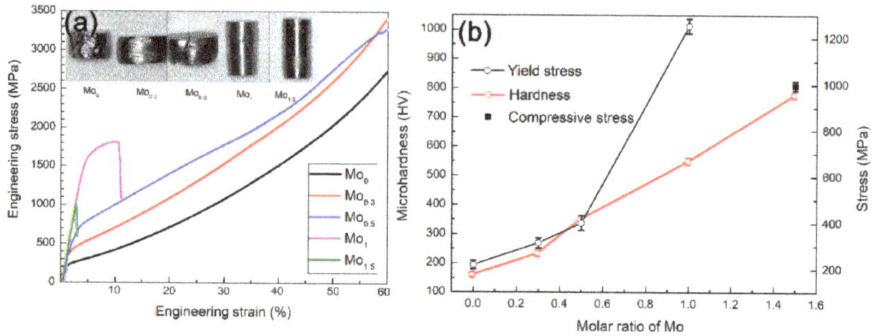

Figure 6. (**a**) Compressive engineering stress-strain curves of Mo*x* high entropy alloys, in which the photos of the cylindrical samples of Mo*x* HEAs after compression tests were inset, and (**b**) Vickers hardness, yield and compressive stress obtained from (**a**).

Figure 7 presents the SEM images of the fractured surface after compressive tests. For Mo1, cracks in the particles are observed, indicating an inter-particle fracture. The FCC solid solution presents a ductile fracture, surrounding the particles. The brittle fracture in the particles becomes more evident for Mo1.5, presenting very little ductile deformation. The crack mainly initiated along grain boundary/precipitate interfaces and large intermetallic particles.

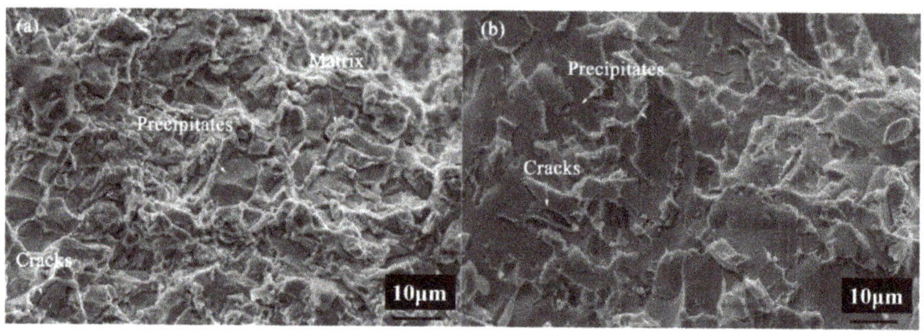

Figure 7. SEM secondary electron images of the fractured surface for (**a**) Mo1 and (**b**) Mo1.5 after compressive tests.

3.4. Tribological Behavior

Figure 8a shows the variation of friction coefficient with sliding distance for the Mo*x* HEAs, and Figure 8b shows the average friction coefficient. It can be observed that the Mo0 and Mo0.3 generate the highest friction coefficient at approximately 0.62, when sliding against YG6 ball, followed by Mo1.5 at 0.45 and Mo0.5 at 0.39, the Mo1 shows the lower friction coefficient at 0.35. The FeCoCrNi alloy is soft and during the wear process, the more debris accumulated on the worn surface maybe increase the roughness of the worn surface and increased the friction coefficient. The increase of the Mo additive increased the hard secondary phase content and reduced the friction coefficients, it can be inferred that the reciprocating wear resistance can be improved. From Figure 8b we can see Mo1 alloy shows the lower average friction coefficient.

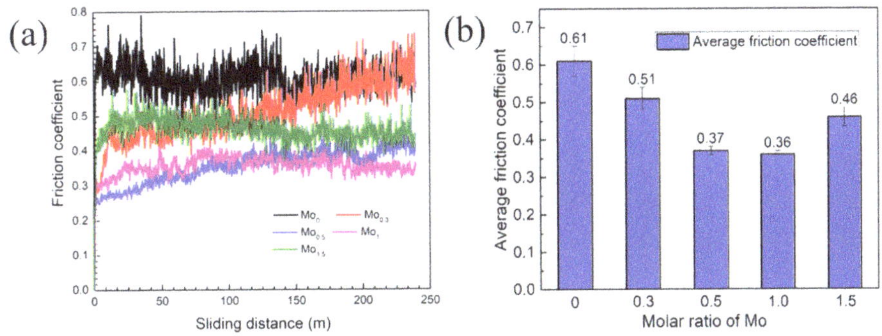

Figure 8. (**a**) Evolution of friction coefficients with the sliding distance for the Mo*x* HEAs in wear process after 240 m sliding under 20 N at a speed of 0.1 m/s, (**b**) the average friction coefficient of Mo*x* alloys.

3.5. *Surface Profilometry*

Figure 9a,b exhibit the 3D profilometry of the wear tracks after 240 m sliding wear test under the load of 20 N at 0.1 m/s. The color change in the figure means the wear track depth change as illustrated in Figure 9b. Mo0 alloy has a maximum depth of 209 μm and a maximum width of 2140 μm. With the increase of Mo molar ratio, both the depth and width of the wear track decreased till the molar ratio of Mo reaches 1.5, whose wear track grows wider and deeper than that of Mo1. This indicates that Mo1 alloy has the lowest depth and width among the tested HEAs, which shows the best wear resistance.

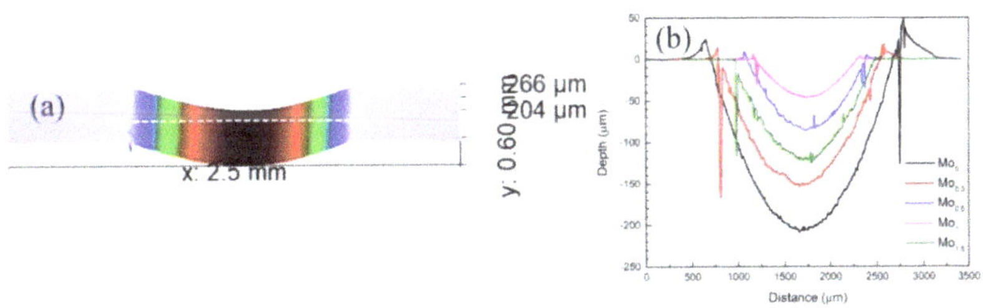

Figure 9. (**a**) Representative 3D profilometry of the wear tracks for Mo1 and (**b**) the measured wear track profiles for the wear tracks of samples with different Mo molar ratios.

Figure 10 shows the variation of volume wear loss, wear rate, and average surface roughness (Sa) with the molar ratio of Mo. The volume loss is obtained by multiplying the cross-sectional area of the wear track by the length of the wear track, the values of which is averaged from at least three wear tests. The wear rate in this paper is calculated according to Archard's law: $K = V/(F \times L)$ [22]. The parameters of K, V, F, and L indicate the wear rate, wear volume (mm^3), applied load (N) and total sliding distance (m), respectively. The trend of the volume loss confirmed the best wear resistance of Mo1 HEA. In addition, to further enrich and accurately explore the wear resistance of the material, we also calculated the wear rate of the Mo*x* HEAs. As presented in Figure 10, the wear rate of Mo*x* HEAs shows the same trend as volume loss and Sa. The wear rate has a rapid decline in Mo*x* alloy, the wear rate of Mo*x* alloy is about 8.5×10^{-5} mm^3/(N·m). When the Mo molar ratio increased up to 1, there is an obvious decreased trend in wear rate, declining to 8.25×10^{-6} mm^3/(N·m). In this paper, the effect of Mo concentration on the wear behavior of Mo*x* HEAs is consistent with previous studies [13,23]. The roughness of the wear scar is consistent with the amount of wear loss and the friction coefficients in Figure 9, which shows that the stronger wear resistance of the material will result in a lower Sa, which then leads to a lower friction coefficient.

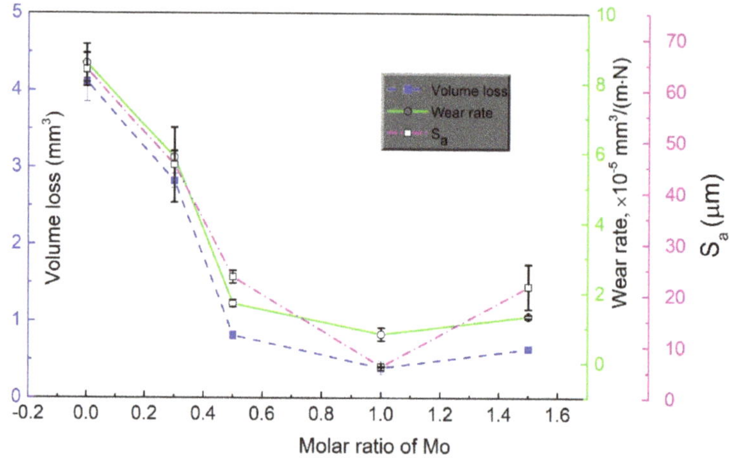

Figure 10. Volume loss, specific wear rates, and the average surface roughness (Sa) of wear scars for Mo*x* (*x* = 0, 0.3, 0.5, 1, 1.5) high entropy alloys.

Figure 11 presents the SEM images of the worn surface for Mo0, Mo0.3, Mo0.5, and Mo1.5 after tribological tests, and Figure 12 shows the SEM images and EDS elemental mapping of Mo1.0. From Figure 11 combined with Figure 12, it can be clearly seen that the width of wear tracks reduces gradually with the Mo molar ratio in the range of 0 to 1, then it increases from the Mo molar ratio of 1 to 1.5 with the Mo1 showing the narrowest wear track. This trend agrees well with the one obtained from the profiles of wear tracks in Figure 9. For the Mo0 HEA, as indicated in Table 3 and Figure 6, it has a low hardness of 3.8 GPa, yield stress of 207.4 MPa and superior ductility. In comparison, the YG6 possesses a superior hardness of 20.1 GPa. Therefore, the hard YG6 asperities will severely plow and deform the soft Mo0, making the Mo0 work-hardened and becomes increasingly brittle, where the maximum shear stress will be generated at a certain depth below the surface [24]. When the accumulation of the plastic deformation at the subsurface exceeds the limit, the cracks will initiate and propagate parallel to the surface, when the crack propagation reaches the surface, it will delaminate from the wear surface [25]. For Mo0.3, the plow grooves can still be observed across the wear track, the wide grooves with the width of 9.2 μm can be seen, which indicates that it was still plowed by the hard YG6 asperities. Some compressed sheets were found to cover some areas of the wear track, indicating the excellent ductility of Mo0.3. Therefore, the Mo0.3 suffers from the dominant micro-cut wear and some adhesive wear. When the Mo molar ratio continues to increase to 0.5, the micro-cut was identified to be the main wear mechanism, some micro-fatigue was also observed with some peeling-off found. Such micro-cut was severely limited for the Mo1 HEA, which suffers from mild abrasive wear. When the molar ratio of Mo increases to 1.5, the wear track becomes wider than that of Mo1 sample and was also distributed by the voids and wear debris, as shown in Figure 11g,h. The formation of these voids is mainly attributed to the micro-fracture mechanism. The hard asperities of YG6 indent the brittle Mo1.5 causing the initiation of the cracks, which will propagate. Such fractures will detach when the cracks propagate to the surface [26].

From Figure 12, it can be seen that a lot of secondary phases were detached from the worn surface, and the FCC solid solution becomes smooth. Interfacial cracks between the FCC solid solution and precipitate are observed, it is attributed to the difference of plastic deformation of the two phases under the interfacial shear stress. Cracks in the precipitate are also found, indicating the brittle fracture of the precipitate under heavy normal stress [24]. It should be noted that the brittle fracture and grain pull-out of the precipitates do not cause the large-scale mechanical damages on the worn surface, they were mainly limited to surface damage of around 10 μm diameter. Because the hard and brittle precipitates were surrounded by the ductile FCC solid solution, the propagation of the cracks into the FCC solid solution was prevented due to the excellent ductility of the FCC solid solution.

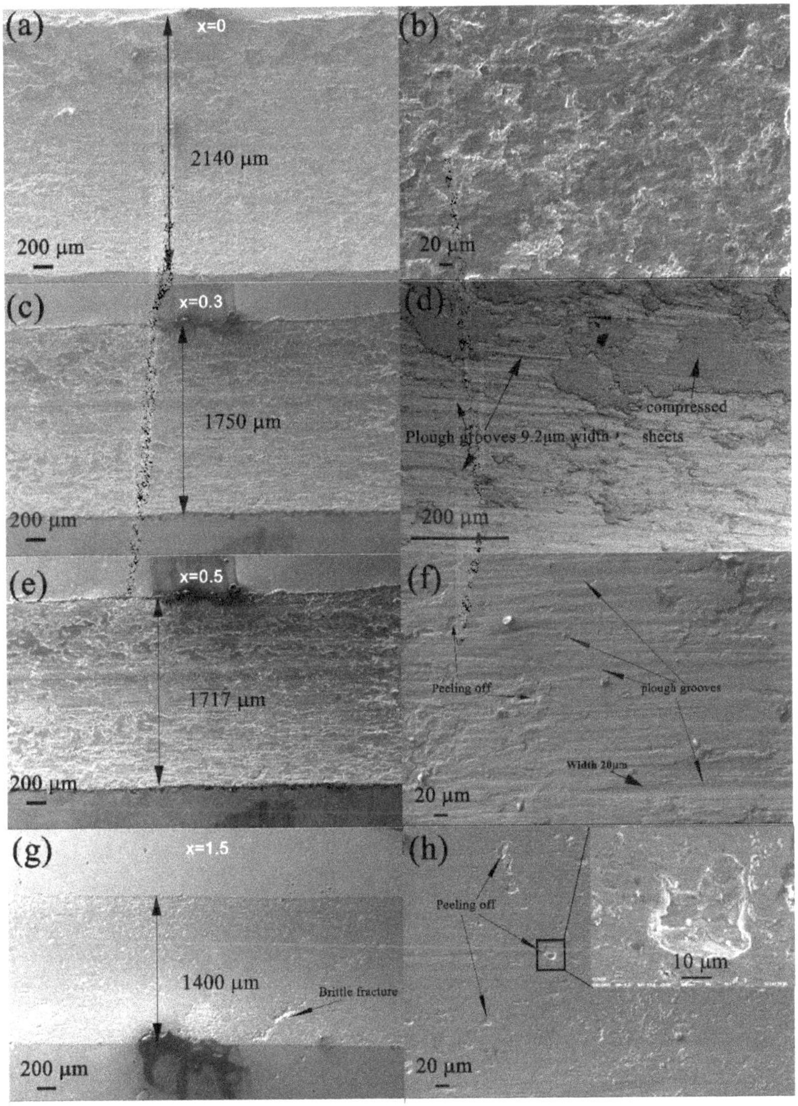

Figure 11. Volume loss, specific wear rates, and the average surface roughness (Sa) of wear scars for Mox (x = 0, 0.3, 0.5, 1, 1.5) high entropy alloys, (**a**) and (**b**) for Mo0 alloy, (**c**) and (**d**) for Mo0.3 alloy, (**e**) and (**f**) for Mo0.5 alloy, (**g**) and (**h**) for Mo1.5.

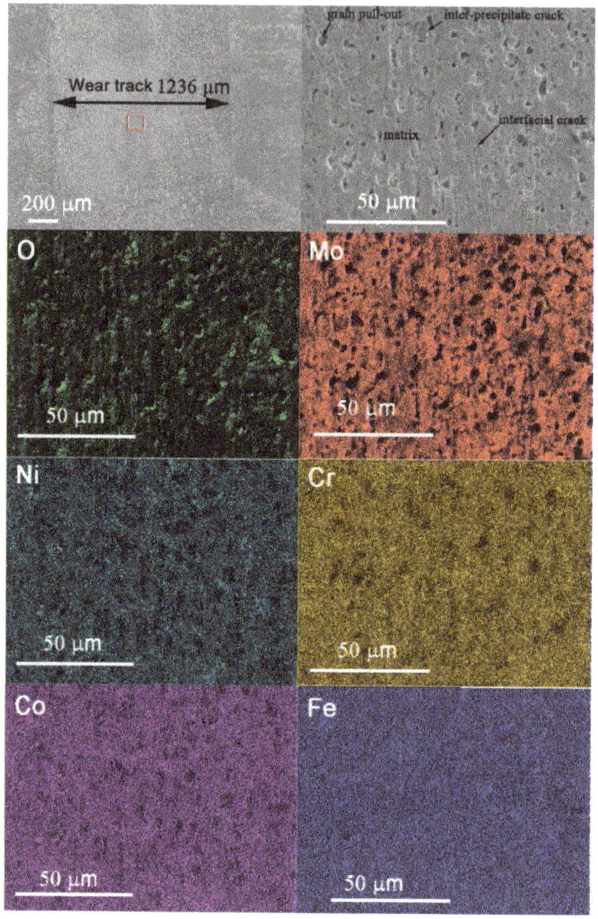

Figure 12. SEM EDS secondary electron images and elemental mapping of the wear track formed on Mo1 HEA after 240 m sliding under 20 N, the red rectangular area was further magnified to its right side.

3.6. Subsurface Characterization

Figure 13 shows the SEM images of the cross-sections of the wear tracks on HEAs, and the work-hardening behavior of the Mo0.3 and Mo0.5 FCC solid solution. As for the HEA without Mo, no obvious protective tribo-oxidation (glaze) layer on the wear track is observed, further indicating the direct YG6 /HEA contact. Such a phenomenon was mainly attributed to the strong oxidation resistance because of the massive Cr atoms. Cracks propagation from the surface to subsurface is observed, indicating an early sign of delamination wear. When the Mo molar ratio was increased to 0.3, a similar contact surface to that of Mo0 was observed, only a few precipitates with diameters less than 3 μm at the top surface are found, whose diameter is pronouncedly smaller than the width (about 9.2 μm) of the plowing grooves, as shown in Figure 11d, indicating the easy removal of these particles through micro-cutting. As for the Mo0.5, the volume fraction of the secondary phases increases slightly, which further reduces the wear rate, as shown in Figure 9. but the amount of the hard phases is still too low, only a small fraction of the top surface was occupied by the hard phase, furthermore, similar problems with the small size of these precipitates is still present, as shown in the inset of Figure 13c. This makes a lot of the precipitates not work effectively to resist the wear. As for Mo1, the cross-sectional SEM image in Figure 13d show that a large area of the contacting surface was occupied by the precipitates, composed of both σ and μ precipitates. This indicates the majority of the load was carried by the precipitates, which have superior elastic modulus and hardness. The diameter of these precipitates is larger than those of Mo0.3 and Mo0.5, which indicates that they cannot be removed off easily. But some pull-outs of the precipitates were observed, the hard precipitation can resist the abrasion of the YG6 asperities, while the ductile FCC solid solution increases the global ductility of the HEA. For Mo1.5, the contacting surface was almost entirely covered with the hard and brittle precipitation, which indicates a major carrier of the load; however, the inter-precipitate cracks can be seen prevalently on the top surface and in the subsurface. The cracked precipitate on the top surface will be removed off easily by the shear stress, causing severe wear of Mo1.5.

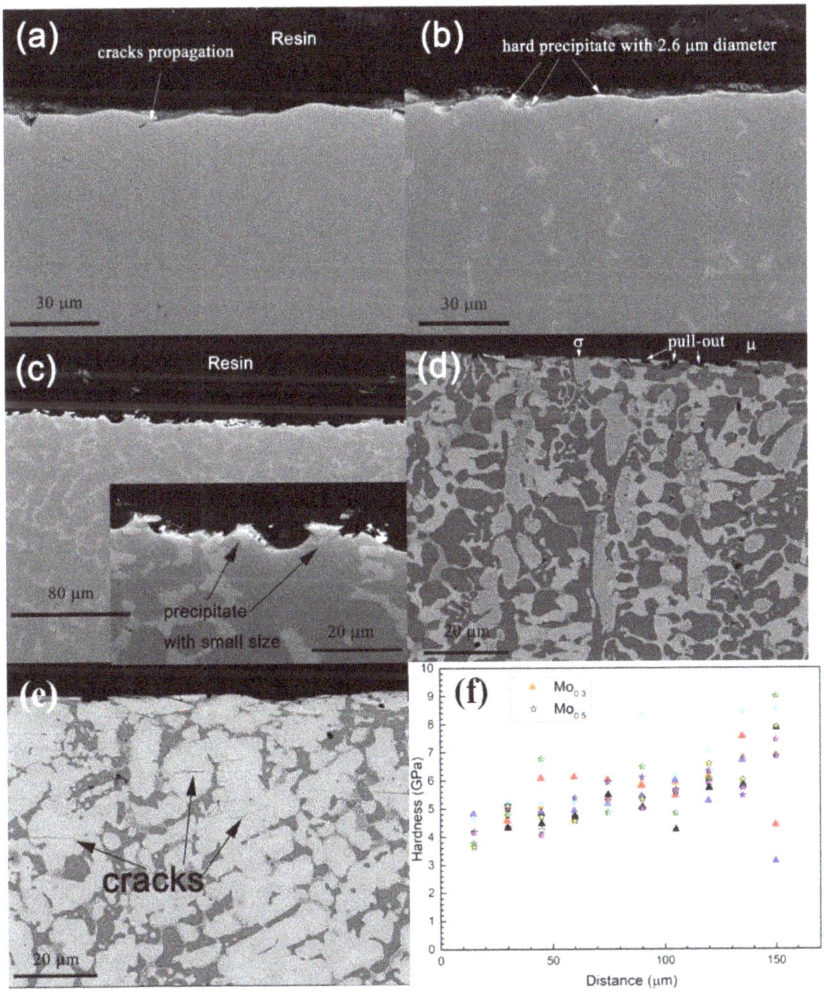

Figure 13. The SEM images of the cross-sections of the wear tracks on high entropy alloys (HEAs), and the work-hardening behavior of the Mo0.3 and Mo0.5 FCC solid solution, (**a**) for Mo0 alloy, (**b**) for Mo0.3 alloy, (**c**) for Mo0.5 alloy, (**d**) for Mo1 alloy, (**e**) for Mo1.5 alloy, (**f**) is the hardness of Mo0.3 and Mo0.5 alloy.

Figure 13f shows the hardness evolution from the top surface to the depth of the subsurface. It can be observed that obvious work hardening occurs, and the hardness of the top surface increases to 7.7 GPa from 4.7 GPa. The Mo0.3 and Mo0.5 experience similar work hardening behavior, indicating that the FCC solid solution undergoes similar plastic deformation. The detailed microstructure of the subsurface was revealed by Deng et al. [19] using FIB and STEM, it found that the subsurface can be divided into three layers: The top layer is composed of very fine nano-grains, the medium layer is the transition region, and the bottom layer belongs to the FCC solid solution. Thus, it can be concluded that the highest hardness is due to the formation of nano-grains. These nano-grains are formed due to the repeated plastic deformation of the FCC solid solution.

4. Discussion

Different from some metals, which can establish the wear-resistant glaze layer through tribo-oxidation, agglomeration, and compaction of the wear debris in the sliding process [27,28]. The tribo-oxidation of Mo*x* HEAs is believed to play a weak role in the wear resistance of Mox HEAs. From Figure 12, it can be postulated that the sliding surface of Mo*x* HEA suffers from weak oxidation. The intensity of the oxygen signal is weak except for some small areas. The cross-sectional SEM images in Figure 13 also demonstrate that no tribo-oxidation layers exist on the top sliding surfaces. This is mainly attributed to the high Cr content in the FCC solid solution and precipitates. Deng et al. analyzed the subsurface using FIB and STEM techniques, though they found that partially tribo-oxidation of wear debris occurs and loosely agglomerated; however, the dominant FCC solid solution was not oxidized [19]. Therefore, the direct sliding contact between hard YG6 and Mo*x* dominates in the entire sliding process, in which the hard precipitates play a critical role in the friction and wear responses.

The global hardness of the Mo*x* HEAs increases continuously with the Mo molar ratio, as shown in Figure 6b, which is beneficial to the improvement of abrasion resistance. The Nanoindentation results in Figure 5 suggest that the FCC solid solution is strengthened when the Mo molar ratio is increased from 0 to 0.3, further increase in Mo will not further strengthen the FCC solid solution due to the dissolution saturation of Mo. It will lead to more precipitation of the hard intermetallic. The increased volume fraction of intermetallic results in increased global hardness. Simultaneously, it will also increase the yield stress as well from Mo0 to Mo1.0. These precipitations can work as obstacles to the movement of dislocations [29], which also can reduce the plastic flow of the soft FCC solid solution. However, it will deteriorate the ductility. Therefore, when the Mo molar ratio is increased to 1.5, the HEA suffers from brittle fracture without any plastic deformation. This is mainly due to the overly high-volume fraction (77.5%) of the intermetallic precipitates, which severely break the continuity of the FCC solid solution, as shown in Figure 7. In this case, the HEA is more prone to micro-fracture when subject to external stress, and the cracks will propagate rapidly after its formation.

The Mo1 is verified to have the lower average friction coefficient and the highest wear resistance for the relative sliding process among the tested HEAs. It is because the Mo1 has the desirable combination of mechanical properties with the global hardness of 550 HV, yield stress of 825 MPa, and a compression rate of 11.4% before fracture. This is mainly attributed to the desirable size, volume fraction, and distribution of the hard-reinforcing precipitation within the ductile FCC solid solution. Many wear-resistant composites enhance their wear resistance by introducing secondary hard phases into the ductile matrix material, such as ceramic particles strengthened high chromium white cast iron composite castings used in mineral and cermet crushing industry [30], Al_2O_3 enhanced aluminum for lightweight and wear-resistant transportation material [31]. It was usually found that the wear resistance firstly improved with the increasing content of hard-secondary phases, then declined when the volume fraction of the secondary phases exceeds a certain value.

Figure 14 presents the schematic diagram of the interactive behaviors between the precipitates and YG6 asperities in the wear process under heavy normal and shear stress.

Firstly, the ductile FCC solid solution encompassing the hard phase can support the hard phase and distribute the normal and shear stress more uniformly, as shown in Figure 11 (asperity a). this could improve the load-bearing capability of the HEAs. The presence of the hard phases also can improve the global hardness to reduce the penetration of the hard asperities or abrasive particles [32], as shown in Figure 11 (asperity b). According to the friction theory suggested by Bowden and Tabor [33], the frictional resistance originated from two parts: One is the force to shear off the metallic junctions formed by adhesion and welding at the interface and the other is the plowing force to displace the softer metal from the path of the harder metal. The YG6 ceramic ball, composed of 94%WC and 6% Co, shows a dissimilar chemical nature from the Mox HEAs, therefore, the role of adhesion can be negligible and the plowing force becomes the dominant factor. With a high fraction of the sliding surface occupied by hard precipitates, which present similar hardness to that of YG6 (Table 3), the penetration of YG6 asperities into the HEAs was sensibly resisted. This can markedly reduce the plowing force (friction force), Furthermore, the structure that the soft FCC solid solution surrounding the hard phase can enable the hard phase to move relatively freely, it could make the hard phase move downward to some extent, when subject to heavy stress, this could reduce the probability of the micro-fracture. However, these precipitates also experience some mechanical damages, mainly due to the low toughness. They will fracture. When the interfacial bonding is weak, the precipitate will also be pull-out by the shear stress (asperity f). these explain the continuous reduction of friction coefficient as well as the wear caused by micro-cutting with the Mo element. However, when the Mo molar ratio is too high, such as the volume fraction of 77.5% in Mo1.5, the contacting surface is almost entirely covered by these brittle intermetallic, as shown in Figure 13e. it changes into a significant hard phase-hard phase interaction. This makes the hard and brittle secondary phase suffer from micro-fracture easily, resulting in a higher friction coefficient and material removal rate. Furthermore, the precipitate also works as hard obstacles to resist the plow by the hard asperities or abrasive particles in the sliding, attributing to the superior hardness of 17 GPa. It could substantially reduce the abrasive wear, especially micro-cutting, which is the dominant wear mechanism for Mo0, Mo0.3, and Mo0. 5. Such claims are consistent with the results obtained by Deng et al. [19], who reveals that Mo0-0.3 exhibited the same wear mechanisms.

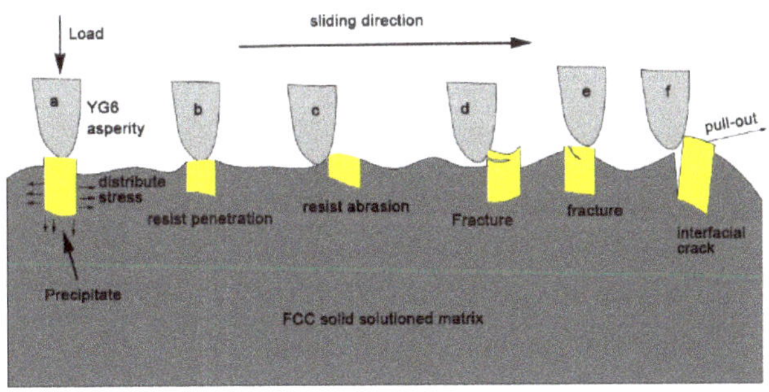

Figure 14. The schematic diagram of the behavior of precipitates in the wear process.

The analysis of the cross-sections reveals that the size of the hard precipitate also plays a key role in the wear resistance. As shown in Figure 11f that the width of the plowing groove is about 20 μm, if the diameter of the precipitate is smaller than 20 μm, the precipitate will be plowed away along with the ductile FCC solid solution, as shown in Figure 13c. Therefore, the diameter of the secondary phase shall be larger than 20 μm to resist the severe abrasion of the counter YG6 asperities, which is achieved for Mo1 HEA. This study explores the effect of Mo element in FeCoCrNiMox high-entropy alloys in the mechanical properties and tribological behavior. It had provided some insights into the tribological behaviors of the HEAs, which are applied in mechanical elements subjecting to relative motion. It will give the designers some enlightenment in the adjustment of chemical composition and microstructure of HEAs to achieve the best wear resistance.

5. Conclusions

The FeCoCrNiMo*x* (x = 0, 0.3, 0.5, 1, 1.5 in molar ratio) high entropy alloys were fabricated by vacuum arc melting, and their microstructure, chemical composition, and mechanical properties were characterized. The friction and wear properties of the prepared alloys with the variation of Mo molar ratio were evaluated. We have explored the differences in performance under different phase structures by regulating the changes in molybdenum content, and have discovered FeCoCrNiMo1 alloy with excellent comprehensive performance in application value. Some conclusions can be obtained as follows:

(1) The as-cast FeCoCrNiMo0 alloy has a simple FCC solid solutioned structure. The volume fraction of precipitates increases continuously with the Mo content, and the maximum value reaches 77.8% for FeCoCrNiMo1.5.

(2) The hardness of the FCC solid solution is strengthened when the Mo molar ratio increases from 0 to 0.3, further increase of the Mo element results in little change of the FCC solid solution hardness. However, the global hardness and yield stress of FeCoCrNiMox increases with the Mo element continuously due to the increased precipitation of the hard intermetallic. But the ductility deteriorates, and the FeCoCrNiMo1.5 HEA does not exhibit any plastic deformation before compressive fracture.

(3) The FeCoCrNiMo1 alloy shows the lower average friction coefficient and wear rate among the tested HEAs, due to the balanced achievement of strength and ductility. When the Mo molar ratio is less than 1, severe micro-cutting and delamination wear was the dominant wear mechanism. It changes to mild micro-cut with some micro-fatigue wear for FeCoCrNiMo1, and transits to severe micro-cut and micro-fatigue wear for FeCoCrNiMo1.5.

(4) When the Mo molar ratio is increased to 1.5, the overly high-volume fraction and brittle nature of μ precipitates are the causes for its higher wear rather than Mo1.

Author Contributions: Conceptualization, S.C. and Y.L.; methodology, Y.X.; formal analysis, S.C.; investigation, Y.X. and X.X.; writing—original draft preparation, Y.X.; writing—review and editing, Y.X., S.C.; and Y.L.; visualization, Y.Y.; supervision, Y.L.; project administration, X.W.; funding acquisition, W.L. All authors have read and agreed to the published version of the manuscript.

Funding: This research was funded by the Young Scientists Fund of the National Natural Science Foundation of China, grant number 51905213, the National Key R&D Program of China grant number 2017YFB0305100, the Science and Technology Planning Project of Guandong Province, China, grant number 2017B090003005, 2017A090905027, 2017B090903005, Natural Science Foundation of Guangdong, grant number 906055014066.

Institutional Review Board Statement: Not applicable.

Informed Consent Statement: Not applicable.

Data Availability Statement: Not available.

Conflicts of Interest: The authors declare no conflict of interest.

Abbreviations

HEAs	High entropy alloys
FCC	Face centered cubic
XRD	X-ray diffraction
SEM	Scanning electron microscope
EDS	Energy diffraction spectrum
TCP	topologically close-packed phases

References

1. Yeh, J.W.; Chen, S.K.; Lin, S.J.; Gan, J.Y.; Chin, T.S.; Shun, T.T.; Tsau, C.H.; Chang, S.Y. Nanostructured High-Entropy Alloys with Multiple Principal Elements: Novel Alloy Design Concepts and Outcomes. *Adv. Eng. Mater.* **2004**, *6*, 299–303. [CrossRef]
2. Cantor, B.; Chang, I.T.H.; Knight, P.; Vincent, A.J.B. Microstructural development in equiatomic multicomponent alloys. *Mater. Sci. Eng. A* **2004**, *375–377*, 213–218. [CrossRef]
3. Tong, C.J.; Chen, M.R.; Yeh, J.W.; Lin, S.J.; Chen, S.K.; Shun, T.T.; Chang, S.Y. Mechanical performance of the AlxCoCrCuFeNi high-entropy alloy system with multi-principal elements. *Metall. Mater. Trans. A* **2005**, *36*, 1263–1271. [CrossRef]
4. Sun, S.J.; Tian, Y.Z.; Lin, H.R.; Lu, S.; Yang, H.J.; Zhang, Z.F. Modulating the prestrain history to optimize strength and ductility in CoCrFeMnNi high-entropy alloy. *Scr. Mater.* **2019**, *163*, 111–115. [CrossRef]
5. Sun, S.J.; Tian, Y.Z.; Lin, H.R.; Yang, H.J.; Dong, X.G.; Wang, Y.H.; Zhang, Z.F. Achieving high ductility in the 1.7 GPa grade CoCrFeMnNi high-entropy alloy at 77 K. *Mater. Sci. Eng. A* **2019**, *740–741*, 336–341. [CrossRef]
6. Chuang, M.H.; Tsai, M.H.; Wang, W.R.; Lin, S.J.; Yeh, J.W. Microstructure and wear behavior of AlxCo1.5CrFeNi1.5Tiy high-entropy alloys. *Acta Mater.* **2011**, *59*, 6308–6317. [CrossRef]
7. Dai, C.D.; Zhao, T.L.; Du, C.W.; Liu, Z.Y.; Zhang, D.W. Effect of molybdenum content on the microstructure and corrosion behavior of FeCoCrNiMox high-entropy alloys. *J. Mater. Sci. Technol.* **2020**, *46*, 64–73. [CrossRef]
8. Ding, Q.; Zhang, Y.; Chen, X.; Fu, X.; Chen, D.; Chen, S.; Gu, L.; Wei, F.; Bei, H.; Gao, Y.; et al. Tuning element distribution, structure and properties by composition in high-entropy alloys. *Nature* **2019**, *574*, 223–227. [CrossRef] [PubMed]
9. Wu, Z.; Bei, H.; Pharr, G.M.; George, E.P. Temperature dependence of the mechanical properties of equiatomic solid solution alloys with face-centered cubic crystal structures. *Acta Mater.* **2014**, *81*, 428–441. [CrossRef]
10. Gludovatz, B.; Hohenwarter, A.; Catoor, D.; Chang, E.H.; George, E.P.; Ritchie, R.O. A fracture-resistant high-entropy alloy for cryogenic applications. *Science* **2014**, *345*, 1153–1158. [CrossRef] [PubMed]
11. Liu, W.H.; Lu, Z.P.; He, J.Y.; Luan, J.H.; Wang, Z.J.; Liu, B.; Liu, Y.; Chen, M.W.; Liu, C.T. Ductile CoCrFeNiMox high entropy alloys strengthened by hard intermetallic phases. *Acta Mater.* **2016**, *116*, 332–342. [CrossRef]
12. Zhu, J.M.; Fu, H.M.; Zhang, H.F.; Wang, A.M.; Li, H.; Hu, Z.Q. Microstructures and compressive properties of multicomponent AlCoCrFeNiMox alloys. *Mater. Sci. Eng. A* **2010**, *527*, 6975–6979. [CrossRef]
13. Miao, J.; Guo, T.; Ren, J.; Zhang, A.; Su, B.; Meng, J. Optimization of mechanical and tribological properties of FCC CrCoNi multi-principal element alloy with Mo addition. *Vacuum* **2018**, *149*, 324–330. [CrossRef]
14. Shun, T.T.; Chang, L.Y.; Shiu, M.H. Microstructure and mechanical properties of multiprincipal component CoCrFeNiMox alloys. *Mater. Charact.* **2012**, *70*, 63–67. [CrossRef]
15. Lyu, Z.; Lee, C.; Wang, S.Y.; Fan, X.; Yeh, J.W.; Liaw, P.K. Effects of constituent elements and fabrication methods on mechanical behavior of high-entropy alloys: A review. *Metall. Mater. Trans. A* **2019**, *50*, 1–28. [CrossRef]
16. Li, W.P.; Wang, X.G.; Liu, B.; Fang, Q.H.; Jiang, C. Fracture mechanisms of a Mo alloyed CoCrFeNi high entropy alloy: In-situ SEM investigation. *Mater. Sci. Eng. A* **2018**, *723*, 79–88. [CrossRef]
17. Niu, Z.Z.; Wang, Y.Z.; Geng, C.; Xu, J.; Wang, Y. Microstructural evolution, mechanical and corrosion behaviors of as-annealed CoCrFeNiMox (x = 0, 0.2, 0.5, 0.8, 1) high entropy alloys. *J. Alloys Compd.* **2020**, *820*, 153273. [CrossRef]
18. Bae, J.W.; Park, J.M.; Moon, J.; Choi, W.M.; Lee, B.J.; Kim, H.S. Effect of μ-precipitates on the microstructure and mechanical properties of non-equiatomic CoCrFeNiMo medium-entropy alloys. *J. Alloys Compd.* **2019**, *781*, 75–83. [CrossRef]
19. Deng, G.; Tieu, A.K.; Su, L.; Wang, P.; Wang, L.; Lan, X.; Cui, S.; Zhu, H. Investigation into reciprocating dry sliding friction and wear properties of bulk CoCrFeNiMo high entropy alloys fabricated by spark plasma sintering and subsequent cold rolling processes: Role of Mo element concentration. *Wear* **2020**, *460–461*, 203440. [CrossRef]
20. Shun, T.T.; Hung, C.H.; Lee, C.F. The effects of secondary elemental Mo or Ti addition in Al0.3CoCrFeNi high-entropy alloy on age hardening at 700°C. *J. Alloys Compd.* **2010**, *495*, 55–58. [CrossRef]
21. Gengxiang, H.; Xun, C.; Yonghua, R. *Fundamentals of Materials Science*; Shanghai Jiao Tong University Press: Shanghai, China, 2000.
22. Archard, J.; Hirst, W. The wear of metals under unlubricated conditions. *Math. Phys. Sci.* **1956**, *236*, 397–410.
23. Deng, G.; Tieu, A.K.; Lan, X.; Su, L.; Wang, L.; Zhu, Q.; Zhu, H. Effects of normal load and velocity on the dry sliding tribological behaviour of CoCrFeNiMo0.2 high entropy alloy. *Tribol. Int.* **2020**, *144*, 106116. [CrossRef]
24. Wen, S.; Huang, P. *Tribology Principle*; Tsinghua University Press: Beijing, China, 2002.
25. Suh, P.N. The delamination theory of wear. *Wear* **1973**, *25*, 111–124. [CrossRef]
26. Stachowiak, G.; Batchelor, A.W. *Engineering Tribology*; Butterworth-Heinemann: Oxford, UK, 2013.
27. Pauschitz, A.; Roy, M.; Franek, F. Mechanisms of sliding wear of metals and alloys at elevated temperatures. *Tribol. Int.* **2008**, *41*, 584–602. [CrossRef]
28. Huttunen-Saarivirta, E.; Kilpi, L.; Hakala, T.J.; Metsäjoki, J.; Ronkainen, H. Insights into the behaviour of tool steel-aluminium alloy tribopair at different temperatures. *Tribol. Int.* **2018**, *119*, 567–584. [CrossRef]
29. Gengxiang, H.; Xun, C.; Yonghua, R. *Fundamentals of Materials Science*, 2nd ed.; Shanghai Jiaotong University Press: Shanghai, China, 2006.
30. Zheng, B.; Li, W.; Tu, X.; Song, S.; Huang, W. Effect of ZTA ceramic particles strengthened high chromium white cast iron on three-body abrasion behavior. *Mater. Res. Exp.* **2019**, *6*, 116581. [CrossRef]
31. Holz, D.; Janssen, R.; Friedrich, K.; Claussen, N. Abrasive wear of ceramic-matrix composites. *J. Eur. Ceram. Soc.* **1989**, *5*, 229–232. [CrossRef]

32. Tabor, D. The physical meaning of indentation and scratch hardness. *Br. J. Appl. Phys.* **1956**, *7*, 159–166. [CrossRef]
33. Bowden, F.P.; Moore, A.J.W.; Tabor, D. The Ploughing and Adhesion of Sliding Metals. *J. Appl. Phys.* **1943**, *14*, 80–91. [CrossRef]

metals

MDPI

Article

Enhanced Mechanical and Tribological Capabilities of a Silicon Aluminum Alloy with an Electroplated Ni–Co–P/Si_3N_4 Composite Coating

Zhijie Li [1,2], Fei Ma [1,2], Dongshan Li [3], Shanhong Wan [3,*], Gewen Yi [3], Guofang Geng [1,2] and Lingyan Guo [1,2]

1 State Key Laboratory of Engine Reliability, Weifang 261001, China; lizhij@weichai.com (Z.L.); mafei01@weichai.com (F.M.); genggf@weichai.com (G.G.); guoly@weichai.com (L.G.)
2 Weichai Power Co., Ltd., Weifang 261001, China
3 State Key Laboratory of Solid Lubrication, Lanzhou Institute of Chemical Physics, Chinese Academy of Sciences, Lanzhou 730000, China; lids@licp.cas.cn (D.L.); gwyi@licp.cas.cn (G.Y.)
* Correspondence: shwan@licp.cas.cn

Abstract: Ni–Co–P/Si_3N_4 composite coatings were fabricated over an aluminum–silicon (Al–Si) substrate using a pulse-current electroplating process, in which the rapid deposition of an intermediate nickel–cobalt layer was used to improve coating adhesion. The microstructure, mechanical, and tribological behaviors of the electroplated Ni–Co–P/Si_3N_4 composite coating were characterized and evaluated. The results revealed that the electroplated Ni–Co–P/Si_3N_4 composite coating primarily consisted of highly crystalline Ni–Co sosoloid and P, and a volumetric concentration of 7.65% Si_3N_4. The electroplated Ni–Co–P/Si_3N_4 composite coating exhibited hardness values almost two times higher than the uncoated Al–Si substrate, which was comparable to hard chrome coatings. Under lubricated and dry sliding conditions, the electroplated Ni–Co–P/Si_3N_4 composite coating showed excellent anti-wear performance. Whether dry or lubricated with PAO and engine oil, the composite coating showed minimum abrasive wear compared to the severe adhesive wear and abrasive wear observed in the Al–Si substrate.

Keywords: silicon–aluminum alloy; electroposition; Ni–Co–P/Si_3N_4 composite coating; mechanical strength; friction and wear

Citation: Li, Z.; Ma, F.; Li, D.; Wan, S.; Yi, G.; Geng, G.; Guo, L. Enhanced Mechanical and Tribological Capabilities of a Silicon Aluminum Alloy with an Electroplated Ni–Co–P/Si_3N_4 Composite Coating. *Metals* **2022**, *12*, 120. https://doi.org/10.3390/met12010120

Academic Editor: Pasquale Cavaliere

Received: 25 November 2021
Accepted: 4 January 2022
Published: 7 January 2022

1. Introduction

Al–Si alloys are widely used in automotive machineries for the replacement of steel and cast iron [1]. They are specifically available as a material for components such as cylinder liners and compressor sliders that rub against ferrous counterparts. However, those components made of aluminum-based material still suffer plastic flow, and thereby delamination and adhesive wear, due to their inferior mechanical capability, which greatly influences the reliability and service life of moving parts [2–4]. Nevertheless, the ever-increasing demands for lightweight materials with extreme in-service conditions has led to the strengthening of the surfaces of Al–Si alloys, with a specific wear resistance and desirable friction in order to inhibit adhesive wear and reduce delamination. Many efforts have been dedicated to modifying the surface of aluminum alloys with electroplated composite materials [5]. The deposition of tough composite coatings using the electroplating technique on aluminum alloys has been recognized as one effective approach for promoting its surface hardening, damage tolerance, and associated wear-resistance when used in moving components.

For tribological applications, the frictional behavior of nickel–phosphorus (Ni–P)-based coatings has been mostly investigated under dry sliding conditions [6]. Additionally, Ni–P based coatings have demonstrated desirable resistance to wear under lubricated conditions, e.g., bio-oil, water, and engine oil [7–9]. In particular, by adding either hardening particles such as tungsten carbide (WC), silicon carbide (SiC), alumina (Al_2O_3), silicon oxide (SiO_2), and diamond, or lubricating ingredients such as tungsten sulfide (WS_2), molybdenum sulfide (MoS_2), poly tetra fluoroethylene (PTFE), and metallic particles, Ni–P-based coatings can possess superior tribological capabilities and thereby increase the damage tolerance and endurance of mechanical parts subjected to tremendous wear and tear, as compared with Ni–P coatings alone [10]. In particular, silicon nitride (Si_3N_4) is often adopted as the hardening phase in the development of Ni–P composite coatings due to its high mechanical strength, high hot hardness, and excellent wear resistance [11]. Furthermore, cobalt (Co) has been introduced as an additional element into the Ni–P matrix, leading to a decrease in grain size and content of Ni–P [12]. The resulting Ni–Co–P composite coatings demonstrated enhanced hardness, low friction, and improved wear resistance [13]. This study aimed to investigate the mechanical and tribological properties of electroplated composite coatings made of a nickel–cobalt–phosphorus (Ni–Co–P) matrix and strengthening silicon nitride (Si_3N_4) particles, in referenced to Al–Si substrates. An electroplated Ni–Co–P/Si_3N_4 composite coating was made using the electroplating process under optimal conditions. As for the deposition of the protective coating over aluminum-related substrates for protection against wear and corrosion, the adhesion between the coating and substrate must be sufficiently high. For Al–Si substrates, this process always displays rapid oxidation, which subsequently undermines the adhesive strength and service reliability of electroplated Ni–Co–P/Si_3N_4 composite coatings over Al–Si alloy substrates. In general, the electroplating of Ni–P coatings onto aluminum substrates involves complex zincate or stannate pretreatments. To address this problem, a trial involving a rapid deposition of a Ni–Co intermediate layer was performed to replace these complex pre-cleaning steps.

2. Materials and Methods

2.1. Coating Deposition

A hypereutectic Al–Si alloy with dimensions of 39 mm × 14 mm × 3.6 mm was applied as the substrate for electroplating the Ni–Co intermediate and Ni–Co–P/Si_3N_4 composite. Prior to depositing the Ni–Co intermediate interlayer over the Al–Si substrate, the silicon–aluminum surface was degreased in a 5% aqueous solution of 50% sodium carbonate (Na_2CO_3) and 50% trisodium phosphate (Na_3PO_4) at 60–70 °C for 5 min, and then subjected to a pickling-activating treatment for 5 min using a chemical etching process in an acidic bath containing 60–90 vol.% phosphoric acid (H_3PO_4), 5–15 vol.% nitric acid (HNO_3) and sodium fluoride (NaF) 140 g/L, and the remainder made up of up to 15 vol.% water. Subsequently, a bright dipping solution of 20% HNO_3 and 10% hydrogen fluoride (HF) was further applied to improve the surface uniformity and conformity, followed by a rinse with cool water.

Following this, the specimen was immersed in the electrolyte to deposit a Ni–Co intermediate layer and a subsequent Ni–Co–P/Si_3N_4 composite layer. In the first plating step, the specimen was immersed in a strike bath consisting of an aqueous solution of 240 g/L nickel chloride ($NiCl_2$) and 60 g/L hydrochloric acid (HCl) [14], where the PH value was 1–2. The bath was heated to 80 °C and the immersion time was about 1 min. A current density of 30 $mA \cdot cm^{-2}$ was applied during electroplating of the Ni–Co intermediate layer.

The Ni–Co-coated parts were rinsed with cold running water and subsequently immersed in a plating Ni–Co–P/Si_3N_4 bath. The electrolyte contained 250 g/L nickel sulfate hexahydrate ($NiSO_4 \cdot 6H_2O$), 40 g/L nickel chloride hexahydrate ($NiCl_2 \cdot 6H_2O$), 30 g/L boron acid (H_3BO_3), 0.2 g/L lauryl sodium sulfate ($C_{12}H_{25}NaO_4S$), 0–120 g/L cobalt sulfate heptahydrate ($CoSO_4 \cdot 7H_2O$), and 20 g/L Si_3N_4 particles, where the PH value was adjusted to be 4–5 by addition of sodium hydroxide. Analytical chemical and de-ionized water were used to prepare the electrolyte. The Si_3N_4 nano-particulates with an averaged diameter of 20 nm were dispersed directly in the electrolyte. After electroplating, the Ni–Co–P/Si_3N_4 deposit was rinsed with flowing water and annealed at 400 °C for 2 h.

During the electroplating process, the bath was magnetically stirred with a stirring rate of 150 revolutions per minute (rpm). A nickel plate of 40 mm × 40 mm was used as the anode, while the Al–Si specimen with the dimensions 39 mm × 14 mm × 3.6 mm was used as the cathode to be plated. An applied current density of 5 $A \cdot dm^{-2}$ and a time of 30 min were set for deposition of the Ni–Co–P/Si_3N_4 composite coating. The electrolyte temperature was kept at 50 ± 5 °C using a water bath. The PH value of the electrolyte was adjusted at 3.5 using either a sodium hydroxide (NaOH) solution or dilute sulfuric acid (H_2SO_4).

2.2. Materials Characterization

A field-emission scanning electron microscope (FE-SEM, FEI Apreo S, USA) was applied to observe the cross-sectional morphologies of the Ni–Co–P/Si_3N_4 composite coatings, and energy-dispersive X-ray spectroscopy (EDS) was used to determine the chemical composition and Si distribution of the Ni–Co–P/Si_3N_4 composite coatings. An electron probe microanalyzer (EPMA-1600, Shimadzu, Japan) was used to quantitatively determine the volumetric concentration of the Si_3N_4 particles in the resulting Ni–Co–P matrix.

The microstructures of the resulting Ni–Co–P/Si_3N_4 composite coatings were determined by X-ray diffraction (XRD) with a Cu source (λ = 0.15418 nm) at 2θ = 30–90° and a scanning rate of 2θ = 5° min^{-1} with a 0.02 step size.

The microhardness of the Ni–Co–P/Si_3N_4 composite coatings was measured with a MH-5-VM Vickers-hardness tester with a diamond pyramid indenter (Shanghai HengYi Science and Technology Co, Ltd., Shanghai, China). The normal load was set to 10, 25 and 50 g, with a duration of 5 s. At least 11 points were measured on the coating specimen and an averaged value of hardness was given. The adhesion strength between the resulting composite coating and substrate was measured by means of a micro-scratching technique (Anton Paar, Switzerland). A Rockwell diamond indenter with a radius of 0.2 mm and a cone angle of 120° was applied. During each measurement, the load was progressively increased to 60 N at a rate of 5 N/m, and the scratch distance was 5 mm. All the scratch testing was conducted in atmospheric conditions.

The wear resistances of the Ni–Co–P/Si_3N_4 composite coatings sliding against a GCr15 steel ball (ϕ 6 mm) were measured on a CSM tribometer with a ball-on-disc configuration (CSM Instruments Inc., Peseux, Switzerland). The sliding conditions involved a reciprocating amplitude of 5 mm, a normal load of 10 N, a frequency of 5 Hz, and a duration of 4000 s. All the frictional tests were performed under dry and lubricated conditions at room temperature in air (a relative humidity 52–56%). Poly-Alfa-Olefine (PAO) and engine oil were used for the lubricated conditions. Poly-Alfa-Olefine (PAO) has a kinematic viscosity of 29.0–31.0 mm^2/S at 40 °C and 5.7–6.0 mm^2/S at 100 °C, respectively, as provided by the Hao Runte Trading Co., Ltd in Beijing, China. The engine oil was 15W-40 CF-4, and was purchased from the Great Wall Lubricant Corporation of China. The kinematic viscosity of the 15W-40 CF-4 engine oil was 112.8 mm^2/S at 40 °C and 15.28 mm^2/S at 100 °C, respectively. The friction coefficient was recorded automatically during the tests.

The anti-wear performance of the Ni–Co–P/Si_3N_4 composite coatings was estimated from the specific wear rate. After the wear tests, the worn morphology was observed using an Optical Microscope (Carl Zeiss AxioLab A1, Jena, Germany). The specific wear rate of the coated and uncoated Al–Si specimens was calculated according to the equation, $W = V/DL$, where W was the wear rate in $mm^3\ N^{-1}m^{-1}$, V was the wear volume loss in mm^3—which was determined by a MicroXAM-800 three-dimensional profilometer (KLA-tencor, USA)—D was the total sliding distance in m, and L was the applied load in N. This profiler was able to measure the two-dimensional profile of the wear tracks; the cross-sectional area (s) of the wear tracks was accordingly figured out from the two-dimensional profile collected. Thus, the material loss V of the wear tracks was calculated according to the equation, $V = s \cdot A$, where A represented a reciprocating amplitude of 5 mm.

3. Results

3.1. Chemical Composition and Microstructure Characterization

Figure 1 illustrates the cross-sectional view of a Ni–Co–P/Si_3N_4 composite coating over a silicon-aluminum substrate. This type of Ni–Co–P/Si_3N_4 composite coating is 16 μm thick in total, as shown in Figure 1a. There appears to be no obvious contrast between the intermediate Ni–Co layer and the upper Ni–Co–P/Si_3N_4 composite layer in Figure 1b. Since the whole coating composition primarily consisted of nickel and cobalt elements, which have similar atomic numbers, the two layers did not differ. A careful view was given to the connection region between the adhesive Ni–Co intermediate and the Al–Si substrate, but no defect-like cracking was visible. This microscopic characterization well confirmed that rapid deposition of the Ni–Co intermediate layer greatly improved the coating adhesion onto the Al–Si substrate.

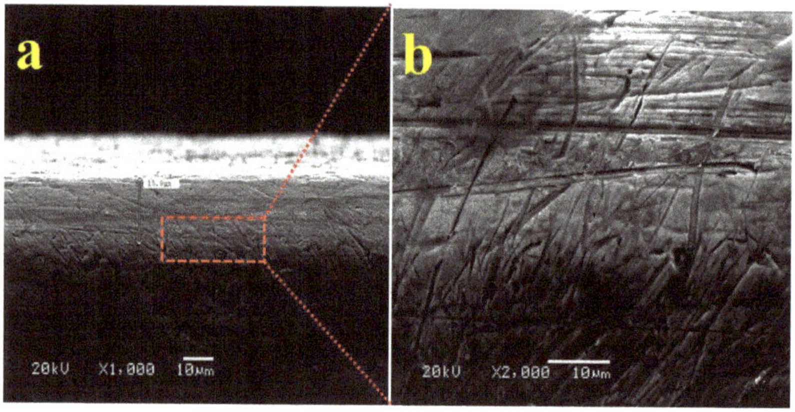

Figure 1. The cross-sectional view (**a**) of a Ni–Co–P/Si_3N_4 composite coating over a silicon aluminum alloy, (**b**) a higher magnification view of selected area in (**a**).

However, the top Ni–Co–P/Si_3N_4 composite layer, intermediate Ni–Co layer, and substrate were not so different from Figure 1. Accordingly, the elemental depth profiling of the Ni–Co–P/Si_3N_4 composite coating over the Al–Si substrate is shown in Figure 2a; above the Al–Si substrate was an intermediate layer of nickel and cobalt that improved the growth and orientation of the subsequently deposited Ni–Co–P/Si_3N_4 composite layer, where the thickness of the intermediate Ni–Co layer was around 5 μm, without any trace of the P element. As for the thickness of the upper Ni–Co–P/Si_3N_4 composite layer, it was about 10 μm thick, as shown in Figure 2a. The EDS analysis, as shown in Figure 2b, shows the elemental composition of the Ni–Co–P/Si_3N_4 composite layer, including 59.26 wt.% Ni, 37.37 wt.% Co, and 3.37 wt.% P. Since the distribution of Si_3N_4 was not homogeneous and its volumetric fraction was minimal within the electroplated Ni–Co–P matrix, the weight concentration of the Si_3N_4 indicated by the SEM/EDS characterization is inaccurate and not applicable for the determination of the exact number of Si_3N_4 particles.

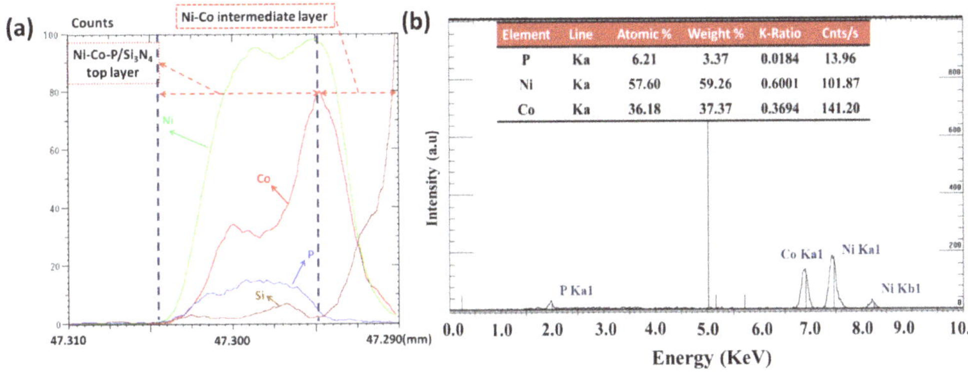

Figure 2. Elemental depth profiling across a Ni–Co–P/Si_3N_4 composite coating over an Al–Si substrate by EPMA (**a**) and elemental composition of a Ni–Co–P/Si_3N_4 composite coating over an Al–Si substrate by SEM/EDS (**b**).

Consequently, EPMA was applied to measure the silicon element within the ROI region of the top layer. As shown in Figure 3a, the silicon mapping was obtained and processed with 1.52a Image J software (National Institutes of Health (NIH). However, as indicated in Figure 3b, an agglomeration of Si_3N_4 particles was present within the resulting Ni–Co–P/Si_3N_4 composite coating. In this study, the nano-sized Si_3N_4 particles were directly dispersed within the electrolyte, assisted by the ultrasonic process prior to electroplating; nevertheless, the electrolyte was just mechanically stirred with the magnetic stirrer during electroplating. Thus, a strong agglomeration trend of Si_3N_4 particles occurred due to their high activity. The volumetric fraction of Si_3N_4 particles is distinguished from the upper Ni–Co–P/Si_3N_4 composite layer in Figure 3b; by assuming that Si_3N_4 was the single hardening phase in the Ni–Co–P matrix, the volumetric fraction of the Si_3N_4 particles was calculated to be about 7.35%.

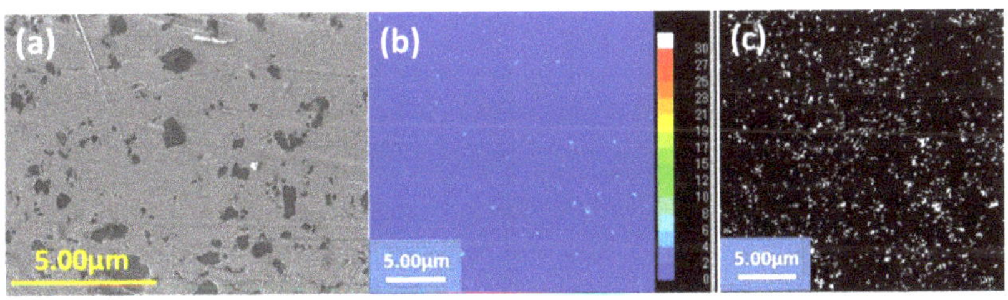

Figure 3. EPMA/EDS mapping of the silicon element within the ROI region (**a**,**b**) and the corresponding volumetric fraction of Si_3N_4 within the Ni–Co–P matrix after Image J processing (**c**).

Figure 4 shows the XRD spectra of three electrodeposited Ni–Co–P/Si_3N_4 composite coatings prepared under the same conditions. This indicates there is no distinct difference in crystalline structure among the resulting three electrodeposited Ni–Co–P/Si_3N_4 composite coatings. In this study, the weight concentration of the cobalt element was around 37.37%; thus, Ni–Co–P/Si_3N_4 composite coatings are commonly regarded as a solution of cobalt and phosphorus alloying elements on a nickel base, which leads to lattice distortion [15–17]. As reported in the literature, the addition of Si_3N_4 to the resulting Ni–Co–P composite coating reduces the hydrogen atoms, leading to defects in the lattice [18]. With the dispersion of Si_3N_4 in the electrolyte, the (111) peak becomes intense, which relates to the change in Co content in the composite coatings. This means the presence of Si_3N_4 particles facilitates the preferential deposition of Co in the composition coating [19]. Those aspects result in the lattice distortion of nickel.

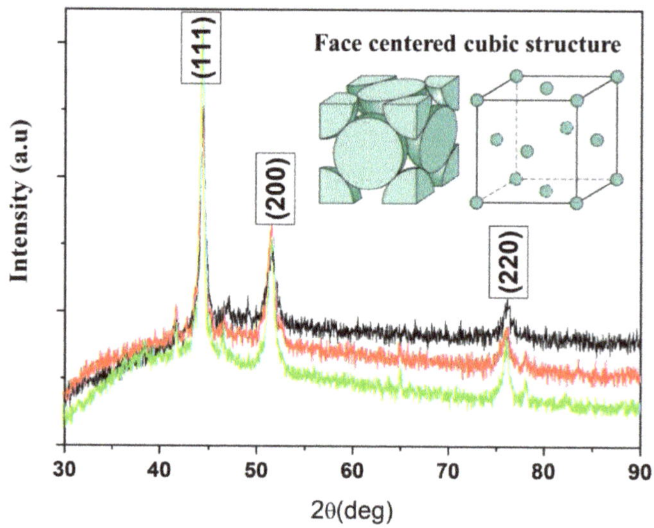

Figure 4. XRD pattern of three Ni–Co–P/Si_3N_4 composite specimens prepared under the same conditions.

As shown in Figure 4, the three representative peaks corresponding to the (111), (200) and (220) crystallographic planes of the Ni–Co solid solution were distinct. Among them, the (111) crystal surface was very intense, which indicates a crystal orientation toward the direction [111] during electroplating, which is assigned to the face-centered cubic (FCC) crystals. In the FCC crystalline system, the surface energy of the (111) plane was lower than that of the (220) plane, the (111) plane had a higher atomic density as compared with those of the other two planes. Therefore, Ni and Co atoms would migrate from the higher energy states of the (220) plane to the lower energy of the (111) plane during annealing; as a result, the production of Ni–Co grains with random crystallographic orientations and textures occurred, which is consistent with previously reported results [20,21]. In the case of the (200) plane, its intensity remained constant—there were no significant changes in the grain growth direction. Additionally, relatively lower diffraction peaks located at 41.7° and 46.6° were observable, which are reportedly assigned to stable a BCT-Ni_3P phase that originates from the dissolution of P into nickel grains, forming an intermetallic Ni_3P precipitation [22]. It has been reported that such precipitation plays a strengthening role in anti-friction and anti-wear [23]. Thus, the resulting composite coating involves a dominant FCC structure mixed with a minor BCT-Ni_3P phase.

3.2. Mechanical Properties

The micro-hardness of the Al–Si alloy and the Ni–Co–P/Si_3N_4 coated specimen was 221 ± 11 HV_{200g} and 473 ± 24 HV_{200g}, respectively. The results clearly reveal that the resulting Ni–Co–P/Si_3N_4 composite coating exhibited a hardness value at least two times higher that of the silicon aluminum substrate, which is comparable with existing data from Ni–Co–P composite coatings [10]. Figure 5 shows the indented traces on the coated specimen under different indentation loads. Ni–Co–P/Si_3N_4 composite coating does not display obvious indentation-induced cracks and collapse. Thus, the Ni–Co–P/Si_3N_4 composite coating is tough and thus sustains a higher loading capacity.

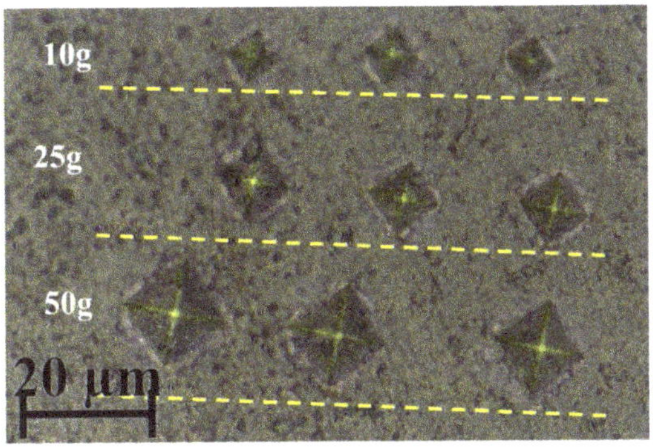

Figure 5. Indentation track of the Ni–Co–P/Si_3N_4 composite coating.

As for the mechanical strength of the Ni–Co–P/Si_3N_4 composite coating, it has been claimed to be due to both the matrix and to strengthening solids [24]. Solid particles generally play a strengthening role, supporting the matrix and resisting deformation. During the electroplating process, a considerable improvement in mechanical strength primarily results from grain nucleation and grain refining, owing to the dispersion of solid particles in metallic matrix [10]. In this study, phosphorus was involved, which resulted in the formation of intermetallic Ni_3P during the annealing process, thereby yielding a hardening effect due to precipitation [25]. Thus, the above-mentioned influential aspects were responsible for the comprehensive mechanical properties of the resulting Ni–Co–P/Si_3N_4 composite coating in this study.

The adhesion strength of the Ni–Co–P/Si_3N_4 composite coating over the Al–Si substrate was measured by micro-scratching testing. As shown in Figure 6a, the adhesion strength was defined as the critical load of 37N for the resulting Ni–Co–P/Si_3N_4 composite coating, where a sudden drop in friction occurred. At this point, the whole coating was scratched off. Figure 6,c show an optical observation of the scratch track. Within the scratch trace, ploughs and deformation phenomenon were present, without observable cracking or fracturing—as shown in Figure 6b. However, some defects are visible in Figure 6c, at a magnification higher than that in Figure 6b. This implies that the densification of the Ni–Co–P/Si_3N_4 composite coating should be further improved.

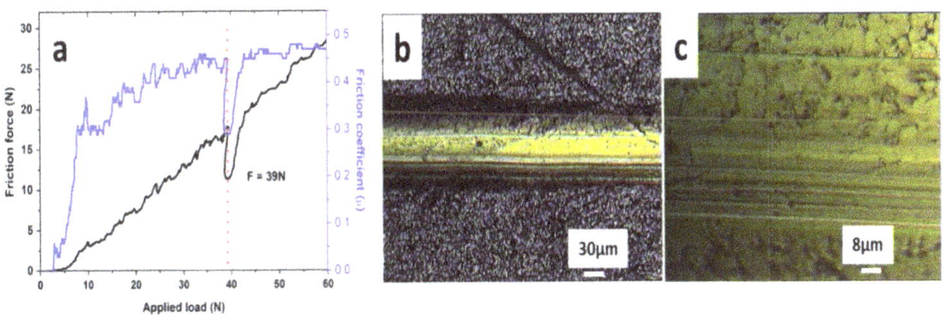

Figure 6. (**a**) The curve of friction and the corresponding friction coefficient versus the applied load during the scratching process, (**b**) Optical images of scratching tracks of the Ni–Co–P/Si_3N_4 composite coating, (**c**) Optical image of the scratching tracks at a higher magnification than (**b**).

3.3. *Tribological Properties of the Ni–Co–P/Si_3N_4 Composite Coating*

The friction and wear behaviors of the Ni–Co–P/Si_3N_4 composite coating under dry and lubricated conditions with PAO and engine oil were investigated to ascertain the suitability of surface hardening for moving components, in reference to the Al–Si substrate. As shown in Figure 7a,b, under dry sliding, friction gradually increased for the Al–Si substrate as sliding occurred, whereas the Ni–Co–P/Si_3N_4 composite coating exhibited unexpected worse friction after running for 25 min—the friction coefficient even rose up to 0.7. When comparing the friction results of the samples lubricated with PAO and engine oil, both the Al–Si substrate and Ni–Co–P/Si_3N_4 specimen demonstrated favorably stable friction, with a friction coefficient of about 0.12.

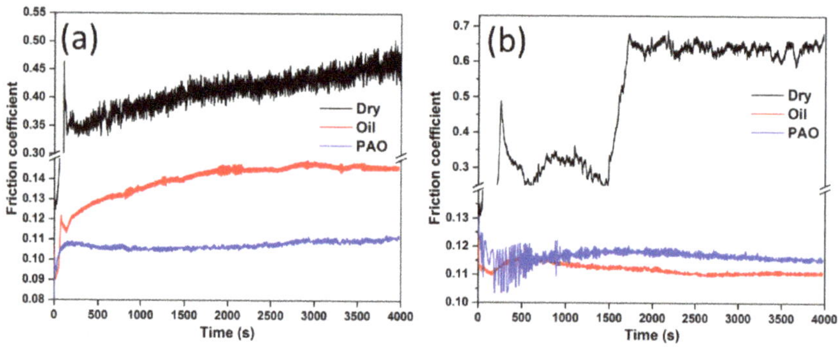

Figure 7. Variation of the friction curves with sliding time for the Al–Si substrate (**a**) and the Ni–Co–P/Si_3N_4 composite coated specimen, under dry conditions and lubrication with PAO and engine oil (**b**).

It is noted that, when the friction pairs were lubricated with engine oil, the friction adversely displayed a considerable increase for the Al–Si substrate in Figure 7a. When adopting PAO as the lubricant, a relative stable friction curve occurred for the Al–Si substrate. This implies that the engine oil involved in this study was not compatible with the Al–Si surface. For the Ni–Co–P/Si_3N_4 composite coating, when applying engine oil as the lubricant, the friction coefficient further decreased, as shown in Figure 7b. Thus, the presence of the Ni–Co–P/Si_3N_4 composite coating favored the reduction of friction.

In this study, a ball-on-flat reciprocating tribometer was adopted to measure the wear loss. Figure 8 displays the specific wear rates of the Al–Si substrate and the Ni–Co–P/Si_3N_4 composite coating under different conditions. It was found that the Ni–Co–P/Si_3N_4 composite coating reduced wear significantly—by two levels of magnitude as compared to the Al–Si substrate. When observing the wear results of the Al–Si substrate specimen under dry and lubricated conditions in Figure 8a, when the PAO was present between the rubbing contacts, a significant reduction of wear occurred. However, the presence of engine oil did not function effectively to inhibit wear, adversely producing a worsening of wear. The material loss of the Ni–Co–P/Si_3N_4 composite coating varied considerably in different situations (Figure 8b), where relatively high wear occurred during dry sliding; PAO performed very well, and applying engine oil caused a slight increase in wear as compared to PAO. However, the role of Si_3N_4 in the tribological behavior of the resulting Ni–Co–P composite coating is not investigated in this study. The existing literature has shown that mixing hardening Si_3N_4 particles into a Ni–Co–P alloyed matrix greatly improves the wear resistance of Ni–Co–P coatings [11,26]. Thus, addition of Si_3N_4 as a hardening phase in alloyed Ni–Co–P coatings protects the Al–Si substrate from wear.

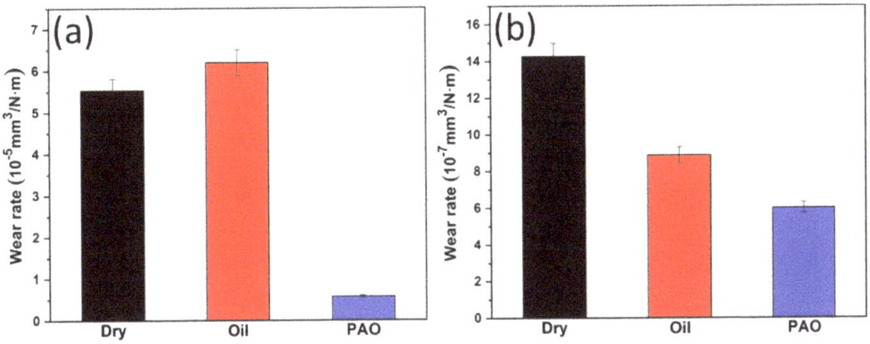

Figure 8. Specific wear rates of Al–Si substrate (**a**) and (**b**) Ni–Co–P/Si_3N_4 composite coated specimens under dry conditions and lubrication with PAO and engine oil.

The optical images in Figure 9 exhibit the topographical features of wear tracks at 100 times zoom, where the dimension of the wear tracks on electroplated Ni–Co–P/Si_3N_4 composite coating surfaces (Figure 9g–i) are observed to be much smaller and shallower than those of the Al–Si substrate surface (Figure 9a–c), whether under dry conditions or lubrication with PAO40 or engine oil. According to the topographical features of the worn Al–Si substrate at higher magnifications of the corresponding wear tracks (Figure 9a–c), the wear mechanisms were severe abrasive wear and deformation, regardless of the lubricating conditions—as shown in Figure 9d–f—which is consistent with the reported tribological characteristics of aluminum alloys in comparison to steel [27]. In comparison, the Ni–Co–P/Si_3N_4 composite coating showed mild abrasive wear, with distinguishable wear tracks parallel to the sliding direction in Figure 9j–l.

According to the specific wear rates in Figure 8, PAO was the best lubricant in this study. When measuring the worn profile, the wear of the Ni–Co–P/Si_3N_4 composite coatings was not readily measured. From the optical observation, some scratch traces are visible in Figure 9h, which is due to the action of an occasional wear particle that became entrapped between the sliding surfaces. Regarding the engine oil, it performed similarly (Figure 9i) between the rubbing contacts.

As shown in Figure 10a, abrasive grooves were visible in the wear scars during dry sliding. Material transfer appeared in the form of scatter debris and was later sheared to the surrounding edges of the wear scar. In Figure 10d, the wear scar featured as grooves and scratches without gross material transfer, typical of abrasion [28]. When applying PAO as the lubrication medium, wear scars showed signs of mild abrasion with slight grooves, with no lump generated (Figure 10b,e). As engine oil lubricated the rubbing contacts, the wear track was characterized by mild abrasion, as shown in Figure 10c,f.

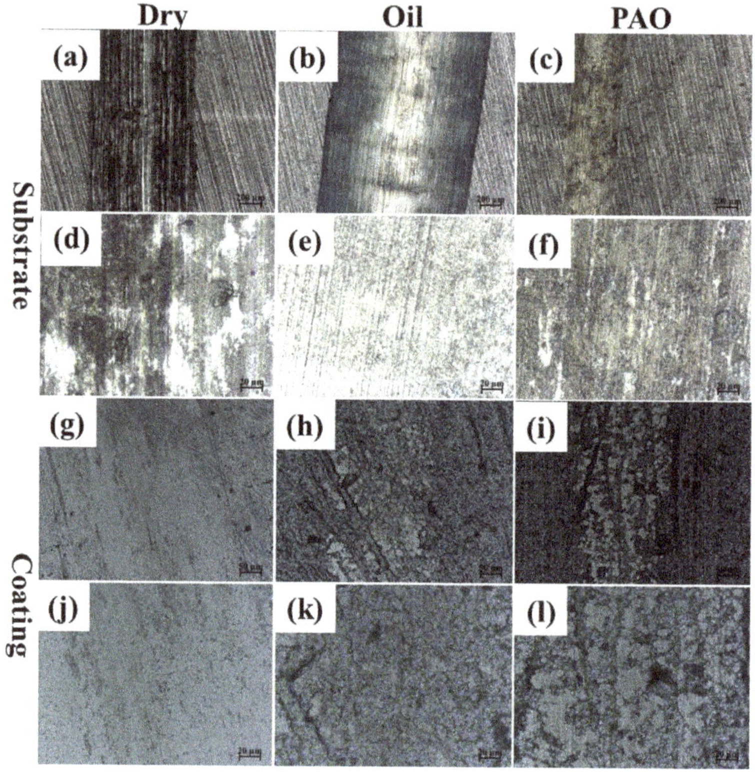

Figure 9. Optical observation of worn traces (**a**–**f**) over Al–Si substrates and (**g**–**l**) Ni–Co–P/Si_3N_4 composite coated specimens under dry conditions and with lubrication with PAO and engine oil.

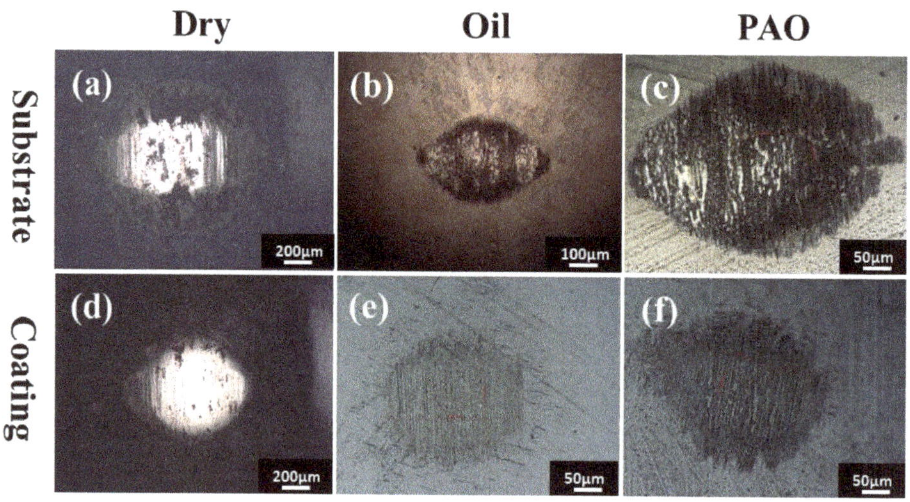

Figure 10. Optical observation of wear scars from GCr15 balls on Al–Si substrates (**a**–**c**) and Ni–Co–P/Si_3N_4 composite coated specimens (**d**–**f**) under dry conditions and with lubrication with PAO and engine oil.

4. Conclusions

The microstructure, mechanical, and tribological behaviors of Ni–Co–P/Si_3N_4 composite coatings prepared using an electroplating process were reported in this study. The rapid deposition of the Ni–Co intermediate layer made a significant contribution towards the sufficiently high adhesion of the resulting Ni–Co–P/Si_3N_4 composite coating over the Al–Si substrate. The presence of a Ni–Co–P/Si_3N_4 composite coating increased the substrate hardness by at least two times. In particular, the Ni–Co–P/Si_3N_4 composite coating demonstrated significant resistance to wear under both dry and lubricated conditions.

Author Contributions: Writing—original draft preparation, Z.L.; formal analysis, F.M.; investigation, D.L. and G.G.; writing—review and editing, S.W.; supervision, F.M. and S.W.; project administration, S.W. and L.G.; funding acquisition, G.Y. All authors have read and agreed to the published version of the manuscript.

Funding: This research was funded by the Science Fund of State Key Laboratory of Engine Reliability (No. SLER-201906), and also received support from the National Natural Science Foundation of China (No. 51575505).

Data Availability Statement: Not applicated.

Conflicts of Interest: The authors have no known competing financial interest or personal relationships that could have appeared to influence the work reported in this paper.

References

1. Tisza, M.; Czinege, I. Comparative study of the application of steels and aluminum in lightweight production of automotive parts. *Int. J. Lightweight Mater. Manuf.* **2018**, *1*, 229–238.
2. Du, F.; Li, C.; Mi, Z.; Shen, Y.; Huang, R.; Han, X.; Dong, Y.; Xu, J. Anti-wear property of aluminum-silicon alloy treated by chemical etching. *Mech. Honing Laser Finish. Mater.* **2019**, *18*, 1273. [CrossRef]
3. Sun, H.; Li, A.; Zhou, Y.; Liao, X.; Ge, D. Dry wear characteristics of machined ZL109 aluminum-silicon alloy surface under unidirectional and reciprocating rolling-contact friction. *Surf. Topogr. Metrol. Prop.* **2019**, *8*, 015001. [CrossRef]
4. Osipov, V.N.; Fadin, Y.A.; Nikanorov, S.P. Wear and Coefficient of Friction of a Supermodified Hypereutectic Aluminum–Silicon Alloy. *Tech. Phys.* **2020**, *65*, 1981–1986. [CrossRef]
5. Zhang, X.; Qin, J.; Das, M.K.; Hao, R.; Zhong, H.; Thueploy, A.; Limpanart, S.; Boonyongmaneerat, Y.; Ma, M.; Liu, R. Co-electrodeposition of hard Ni-W/diamond nanocomposite coatings. *Sci. Rep.* **2016**, *6*, 22285. [CrossRef]
6. Kundu, S.; Das, S.K.; Sahoo, P. Friction and wear behavior of electroless Ni-P-W coating exposed to elevated temperature. *Surf. Interfaces* **2019**, *14*, 192–207. [CrossRef]
7. Duari, S.; Mukhopadhyay, A.; Barman, T.K.; Sahoo, P. Study of wear and friction of chemically deposited Ni-P-W coating under dry and lubricated condition. *Surf. Interfaces* **2017**, *6*, 177–189. [CrossRef]
8. Xu, Y.; Zheng, X.; Hu, X.; Yin, Y.; Lei, T. Preparation of the electroless Ni–P and Ni–Cu–P coatings on engine cylinder and their tribological behaviors under bio-oil lubricated conditions. *Surf. Coat. Technol.* **2014**, *258*, 790–796. [CrossRef]
9. Liew, K.; Chia, S.; Kok, C.; Low, K. Evaluation on tribological design coatings of Al2O3, Ni–P–PTFE and MoS2 on aluminium alloy 7075 under oil lubrication. *Mater. Des.* **2013**, *48*, 77–84. [CrossRef]
10. Lelevic, A.; Walsh, F.C. Electrodeposition of Ni P composite coatings: A review. *Surf. Coat. Technol.* **2019**, *378*, 124803. [CrossRef]
11. Matik, U. Structural and wear properties of heat-treated electroless Ni-P alloy and Ni-P-Si3N4 composite coatings on iron based PM compacts. *Surf. Coat. Technol.* **2016**, *302*, 528–534. [CrossRef]
12. Sarkar, S.; Baranwal, R.K.; Lamichaney, S.; De, J.; Majumdar, G. Optimization of electroless Ni-Co-P coating with hardness as response parameters: A computational approach. *J. Tribol.* **2018**, *18*, 81–96.
13. Shi, L.T.; Hu, J.; Fang, L.; Wu, F.; Xiao, X.L.; Meng, F.M. Effects of cobalt content on mechanical and corrosion properties of electroless Ni-Co-P/TiN nanocomposite coatings. *Mater. Corros.* **2016**, *67*, 1034–1041. [CrossRef]
14. Mallory, G.O.; Hajdu, J.B. *Electroless Plating, American Electroplaters and Surface Finishers Society, Orlando, Fla;* AESF: New York, NY, USA, 1990; pp. 196–201.
15. Ma, C.; Wang, S.C.; Wang, L.P.; Walsh, F.C.; Wood, R.J.K. The electrodeposition and characterization of low-friction and wear-resistant Co-Ni-P coatings. *Surf. Coat. Technol.* **2013**, *235*, 495–505. [CrossRef]
16. Zhang, Y.; Kang, M.; Yao, L.; Mbugua, N.S.; Jin, M.; Zhu, J. Study on the Wear and Seawater Corrosion Resistance of Ni–Co–P Alloy Coatings with Jet Electrodeposition in Different Jet Voltages and Temperatures of Plating Solution. *Coatings* **2020**, *10*, 639. [CrossRef]
17. Ren, L.; Cheng, Y.; Wang, Q.; Yang, J. Study on the properties of Ni-W-P coating with PTFE co-deposition. *Surf. Topogr. Metrol. Prop.* **2019**, *7*, 045009. [CrossRef]

18. Humam, S.B.; Gyawali, G.; Joshi, B.; Kim, T.H.; Lee, S.W. Influence of WC and TaC particles on the microstructure and scratch resistance of electrodeposited nickel-tungsten alloy. *J. Alloys Compd.* **2021**, *893*, 162371. [CrossRef]
19. Jiang, W.; Shen, L.D.; Xu, M.Y.; Wang, Z.W.; Tian, Z.J. Mechanical properties and corrosion resistance of Ni-Co-SiC composite coatings by magnetic field-induced jet electrodeposition. *J. Alloys Compd.* **2019**, *791*, 847–855. [CrossRef]
20. Arora, S.; Kumari, N.; Srivastava, C. Microstructure and corrosion behavior of NiCo-Carbon nanotube composite coatings. *J. Alloys Compd.* **2019**, *801*, 449–459. [CrossRef]
21. Karslioglu, R.; Akbulut, H. Comparison microstructure and sliding wear properties of nickel–cobalt/CNT composite coatings by DC, PC and PRC current electrodeposition. *Appl. Surf. Sci.* **2015**, *353*, 615–627. [CrossRef]
22. Aliofkhazraei, M.; Aliofkhazraei, M.; Walsh, F.C. A review of electrodeposited Ni-Co alloy and composite coatings: Microstructure, properties and applications. *Surf. Coat. Technol.* **2019**, *372*, 463–498. [CrossRef]
23. Karimzadeh, A.; Rouhaghdam, A.S.; Aliofkhazraei, M.; Miresmaeili, R. Sliding wear behavior of Ni–Co–P multilayer coatings electrodeposited by pulse reverse method. *Tribol. Int.* **2020**, *141*, 105914. [CrossRef]
24. Ghavidel, N.; Allahkaram, S.R.; Naderi, R.; Barzegar, M.; Bakhshandeh, H. Corrosion and wear behavior of an electroless Ni-P/nano-SiC coating on AZ31 Mg alloy obtained through environmentally-friendly conversion coating. *Surf. Coat. Technol.* **2020**, *382*, 125156. [CrossRef]
25. Wang, Q.; Callisti, M.; Greer, J.; McKay, B.; Milickovic, T.K.; Zoikis-Karathanasis, A.; Deligkiozi, I.; Polcar, T. Effect of annealing temperature on microstructure, mechanical and tribological properties of nano-SiC reinforced Ni-P coatings. *Wear* **2016**, *356–357*, 86–93. [CrossRef]
26. Dhakal, D.R.; Kshetri, Y.K.; Gyawali, G.; Kim, T.-H.; Choi, J.-H.; Lee, S.W. Understanding the effect of Si3N4 nanoparticles on wear resistance behavior of electroless Nickel-Phosphorus coating through structural investigation. *Appl. Surf. Sci.* **2020**, *541*, 148403. [CrossRef]
27. Decrozant-Triquenaux, J.; Pelcastre, L.; Prakash, B.; Hardell, J. Influence of lubrication, tool steel composition, and topography on the high temperature tribological behavior of aluminum. *Friction* **2021**, *9*, 155–168. [CrossRef]
28. Peng, Y.; Xu, Y.; Geng, J.; Dearn, K.D.; Hu, X. Tribological assessment of coated piston ring-cylinder liner contacts under bio-oil lubricated conditions. *Tribol. Int.* **2017**, *107*, 283–293. [CrossRef]

metals

Article

Tribological Properties of Different-Sized Black Phosphorus Nanosheets as Water-Based Lubrication Additives for Steel/Titanium Alloy Wear Contact

Shaowen Dong [1], Wei Wang [1,*], Yuan Gao [1] and Guanyu Deng [2,*]

1 School of Metallurgy Engineering, Xi'an University of Architecture and Technology, Xi'an 710055, China; 15399057390@163.com (S.D.); yuan-gao@xauat.edu.cn (Y.G.)
2 School of Mechanical, Materials, Mechatronic and Biomedical Engineering, University of Wollongong, Wollongong, NSW 2522, Australia
* Correspondence: gackmol@163.com (W.W.); gd577@uowmail.edu.au (G.D.)

Abstract: Titanium alloys are extensively used in the aerospace, chemical, and biomedical industries. However, it has always been a challenge in the manufacturing and machining of titanium alloys because they exhibit poor friction and wear characteristics, which results in serious problems and significantly restricts their further production and application. Therefore, in the present study, the wear contact between GCr15 steel and Ti6Al4V alloy is specifically studied by considering black phosphorus nanosheets (BP-NS) as water-based lubrication additives, which is expected to have a great potential application in manufacturing and machining titanium alloys. The influence of BP nanosheet size on the coefficient of friction (COF) and wear rate of Ti6Al4V alloy has been comprehensively studied, based on comparisons among adding large BP nanosheets (L-BP) (2–4 μm), medium BP nanosheets (M-BP) (300–500 nm), and black phosphorus quantum dots (BPQDs) (6–10 nm). Compared with ultrapure water, the COF and wear rate of Ti6Al4V alloy are reduced by 42.4% and 82.3%, respectively, when BPQDs are used as water-based lubrication additives. This paper also shows that a lower COF and wear rate is achieved with the addition of BPQDs than the other two BP nanosheet sizes. Derived from the friction tests and worn surface analysis of Ti6Al4V alloy, lubrication mechanisms of different-sized BP lubricants were proposed. The interlaminar shearing between BP-NS and the adsorbed films were the main mechanisms for L-BP and M-BP lubricants, while the adsorption, repair, and ball-bearing effects were mainly presented in the BPQD lubricants. The discoveries in this paper would be beneficial to developing novel lubricants for the manufacturing and machining of titanium alloys.

Keywords: titanium alloys; manufacturing and machining; black phosphorus; lubrication additives; tribological properties

Citation: Dong, S.; Wang, W.; Gao, Y.; Deng, G. Tribological Properties of Different-Sized Black Phosphorus Nanosheets as Water-Based Lubrication Additives for Steel/Titanium Alloy Wear Contact. *Metals* **2022**, *12*, 288. https://doi.org/10.3390/met12020288

Academic Editor: Michael M. Khonsari

Received: 3 January 2022
Accepted: 30 January 2022
Published: 7 February 2022

1. Introduction

Titanium alloys are extensively used in aerospace, chemical, and military industries due to their high strength, favorable corrosion resistance, and better high-temperature properties [1–4]. Nevertheless, titanium and titanium alloy are a typical difficult-to-cut metal, due to high processing temperature and chemical activity, low deformation coefficient, and elastic modulus during processing, which faults limited further application of titanium alloy in the precision parts of the aviation industry [5–7]. Therefore, it is very important to enhance the workability of titanium and titanium alloys. To solve the problems of titanium alloy machining, adding lubrication additives are often used. In view of the difficulties of titanium alloy processing, traditional metalworking fluids are difficult to lubricate effectively during the processing process of titanium alloy [8–11]. In recent years, water-based lubricants have been broadly applied in the metalworking due to their stable chemical properties, high extreme pressure, anti-wear properties, and good cooling capacity. Several studies have shown that adding nanomaterials into water is an effectively methods to improve the lubrication properties [12–14].

The special physical and chemical properties of the nanomaterials have attracted widespread attention in lubrication additives fields [15–17]. The nanomaterials dispersed into the lubrication fluid could enter into the frictional area and form the lubrication film on the surface of metals due to their small sizes [18,19]. Black Phosphorus (BP), as an emerging two-dimensional material, is often used in optical, electrical, and mechanical fields [20,21]. At present, in the area of macroscopic and microscopic tribology, the tribological behaviors of BP-NS as lubrication materials have also attracted the attention of researchers. In our previous investigations, compared with traditional two-dimensional materials (Graphene, Molybdenum Disulfide), BP-NS as n-Hexadecane (16C)-based lubrication additives presented excellent wear resistance [22]. Wang et al. used NaOH-modified BP-NS as a lubrication additive and achieved super-lubricity. The reason is that during the sliding process, the BP-OH nanosheets in the silica gel layer formed by hydrolysis reactions, because the water layer is attached to the BP-OH nanosheets through hydrogen bonds, thus obtaining strong super-lubricity [23]. Wu et al. [24] clearly observed that the frictional force at the degradation area of BP-NS was reduced by about 50% with an atomic force microscope. BPQDs, another zero-dimensional nanomaterial of BP, as water-based lubrication additives, could demonstrate tremendous potential due to the uniform size and good water dispersion. The BPQDs prepared by Ren can be stably dispersed in an aqueous solution of glycol, while achieving macroscopic super-lubricity at a contact pressure of 336 mpa (μ, 0.002) [25]. Tang et al. [26] added BPQDs to triethanolamine aqueous solutions, and the results showed that BPQDs lubrication additives exhibited significant anti-friction and anti-wear properties even at ultra-low concentrations (0.005 wt.%). Seminal works have demonstrated that BP-NS and BPQDs have excellent lubrication performance as lubrication additives.

The focus of this research is the tribological properties of different-sized BP-NS as lubrication additives for the GCr15/Ti6Al4V alloy contact. The BP-NS were thoroughly analyzed for morphology, structure and composition using microscopic and spectroscopic techniques. Lubrication mechanisms were discussed and analyzed by tribological tests and wear surface analysis. This study describes the potential novel applications for different-sized BP-NS as water-based lubrication additives.

2. Experimental

2.1. Materials

Red phosphorus (RP) (99.999%, Guangzhou Chemical Reagent Factory, Guangzhou, China), N-methylpyrrolidone (NMP) ($\geq$99.5%, Aladdin reagent factory, Shanghai, China), and anhydrous ethanol ($\geq$99.7%) were used as raw materials. All reagents did not require further purification unless otherwise stated. Ultrapure water with an electrical resistivity of about 18.25 MΩ·cm was used.

2.2. Preparation of Different-Sized BP-NS as Lubrication Additives

As shown in Figure 1, different-sized BP-NS as lubrication additives were prepared by high-energy ball milling (HEBM) and sample exfoliation methods. The mass ratio of Gcr15 ball (Φ 10 mm, Φ 6 mm) to red phosphorus, ball milling time, and rotation speed are 50:1, 1 h, and 1000 rpm. The final BP powders were collected and stored in an argon glove box.

The preparation method of L-BP is as follows. The BP powders were dispersed into NMP and sonicated in an ice bath for 8 h. The ultrasonic black solutions were centrifuged at 3500 rpm for 15 min and collected. The obtained precipitates were washed three times using alcohol and ultrapure water. Similarly, the black solutions were centrifuged at 3000 rpm to obtain the supernatant and then centrifuged at 7500 rpm for 25 min to obtained M-BP.

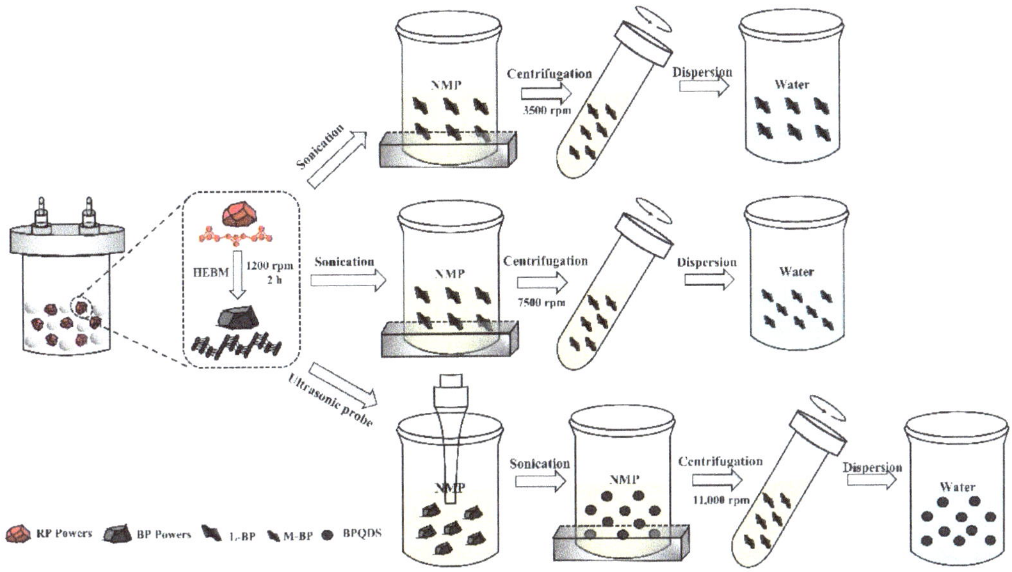

Figure 1. Synthesis process of different-sized BP-NS as lubrication additives.

The preparation method of BPQDs is as follows. The BP powders were dispersed into NMP and sonicated with a probe sonicator for 3 h, followed by ice bath sonication for 8 h. The black solutions were centrifuged at 3500 rpm for 15 min and 7500 rpm for 25 min to obtain the supernatant, and finally centrifuged at 11,000 rpm for 30 min to collect the precipitation. After washing precipitates, the BPQDs were obtained by the vacuum freeze dryer.

The obtained L-BP, M-BP, and BPQDs powder were dispersed into the ultrapure water with the concentrations of 0.025, 0.05, 0.075, and 0.1 wt.% by ultrasonication for 1 h.

2.3. Characterization

The phase structures of L-BP, M-BP, and BPQDs were analyzed by X-ray diffraction (XRD, Bruker D8) (BRUKER, Billerica, USA) and a Raman spectrometer (Raman, Horiba Scientific LabRAM HR Evolution) (HORIBA Scientific, Kyoto, Japan). A scanning electron microscope (SEM, Zeiss Sigma 300) (Zeiss, Oberkochen, German) with an energy dispersive spectrometer (EDS) (Zeiss, Oberkochen, German) were used to observe the micromorphology and size distribution of the BP and RP powders. The microscopic morphologies of BPQDs, L-BP, and M-BP were further characterized by a transmission electron microscope (TEM, FEI Talos F200i) (FEI Company, Hillsboro, OR, USA). The wear surfaces of the steel ball and Ti6Al4V disc were analyzed by SEM and EDS. The wear scar profile and wear volume were measured with the 3D scanning profiler. Finally, the element composition of wear track is examined by Raman scattering and an X-ray photoelectron spectroscopy (XPS, PHI). (ULVAC-PHI, Kyoto, Japan)

2.4. Tribological Tests

The tribological properties of GCr15/Ti6Al4V with ultrapure water and the lubrication additives were studied by ball-on-disc tribo-tester (UMT-5, CETR Corporation) at room temperature. The upper sample is a GCr15 steel ball (Φ 6 mm) and the surface roughness (Ra) is about 18.5 nm. The lower sample is the Ti6Al4V disc (ø 24 mm $\times$ 7 mm). The samples were standard samples and the chemical composition is shown in Tables 1 and 2.

Table 1. The chemical composition of the Ti6Al4V disc.

Component	Al	V	Fe	Ti
Percentage (wt.%)	6.59	4.77	0.035	Bal

Table 2. The chemical composition of the Gcr15 ball.

Component	C	Si	Mn	Cr	Fe
Percentage (wt.%)	0.95–1.05	0.15–0.35	0.25–0.4	1.40–1.65	Bal

The tribological tests were performed for 30 min. The radius of rotation, the rotation speed, and the loads were 3.5 mm, 150 r/min, and 8 N–15 N, respectively. All tests were duplicated five times under each condition to confirm the reproducibility of the results. After tests, the ball and disc were cleaned and dried. The wear rate is calculated using this formula:

$$V = \left(\frac{\pi l}{6}\right)\left(\frac{3d^2}{4} + l^2\right) \quad (1)$$

$$l = r - \sqrt{r^2 - \frac{d^2}{4}} \quad (2)$$

$$W = \frac{V}{P \cdot S} \quad (3)$$

where W, V (mm^3), S (m), r (mm), d (mm), l (mm), and P (N) represent the wear rate, wear volume, sliding distance, ball radius (3 mm), wear diameter, wear height, and load of the balls, respectively. The wear diameter is measured by SEM.

3. Results and Discussion

3.1. Characterization and Analysis of the Initial BP Powders

XRD patterns of RP and BP powders are shown in Figure 2a. The RP powders have two characteristic peaks at 2θ = 30 and 55°, corresponding to the amorphous RP. After HEBM for 2 h, two broad characteristic peaks were changed into 2θ = 26.5° (020), 35.0° (021), and 56.0° (111) diffraction peaks, indicating that the BP powder has a standard orthogonal structure (JCPDS Card NO.76-1957) and RP has been converted to BP by HEBM process. In addition, the results of Raman spectroscopy proved the transfer of RP to BP powder (Figure 2b). After the 2 h HEBM process, the characteristic peak of RP at 305 cm^{-1} disappeared and three sharp characteristic peaks appeared in BP, corresponding to the $A^1{}_g$, $B^2{}_g$, and $A^2{}_g$ vibrational modes. Figure 2c reflects that the size of the prepared BP powder is between 200 nm and 3 μm. According to EDS analysis, the mass ratio of P is 95.96 wt.% and the mass ratio of O is 4.04 wt.%; moreover, it can be seen from the EDS analysis that the brown image has a lot of P element content, while the oxygen element content is less, indicating that the prepared black phosphorus has a lower degree of oxidation.

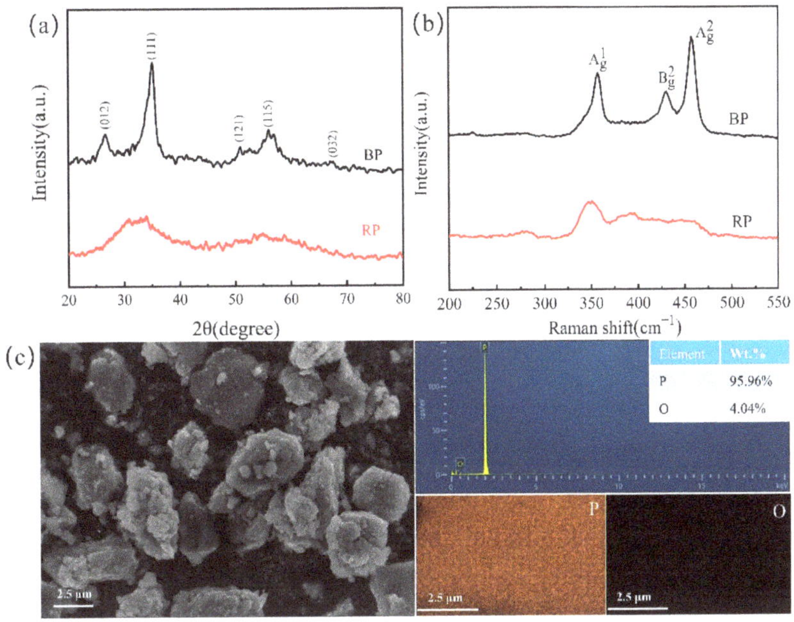

Figure 2. (a) XRD and (b) Raman patterns of RP and BP. (c) SEM image and EDS of BP powders.

3.2. Characterization of Different-Sized BP-NS

The characteristics of L-BP, M-BP, and BPQDs are shown in Figure 3, which indicted that XRD patterns of different-sized BP-NS correspond to the characteristic peaks of the BP powders, and these results are similar to BP orthorhombic crystals (JCPDS No. 74-1878). The diffractive peaks of the L-BP, M-BP, and BPQDs are clearer than that of the BP powders. The blue shift occurred on the (111) plane, indicating that different-sized BP-NS have better crystallinity. The figure of the Raman spectrum (Figure 3c) reveals that the diffractive peaks correspond well to the characteristic peaks of $A^1{}_g$, $B^2{}_g$, and $A^2{}_g$ in the BP powders. Compared with the diffractive peaks of the BP powders, the diffractive peaks of L-BP, M-BP, and BPQDs have a significant blue shift due to mini-size and functional groups. Especially for the BPQDs samples, the characteristic peaks of $A^1{}_g$, $B^2{}_g$ and $A^2{}_g$ are blue-shifted by 1.9, 3.4, and 2.1 cm^{-1}, respectively.

To further analyze the chemical composition and states of the different-sized BP-NS, FTIR spectroscopic were performed. As shown in Figure 3d, the characteristic peaks of 2908, 1637, 1153, and 1002 cm^{-1} of different-sized BP-NS can ascribe the stretching vibrations of -OH, C=O, P=O, and P-O. The above peaks are the absorption of H_2O, CO_2, and the surface oxidation of BP samples. Due to the smaller volume and the larger volume of functional groups of BPQDs, the intensity of the peaks of -OH and C=O in the spectra of BPQDs are much sharper than that in the spectra of L-BP and M-BP. The abundant oxygen-containing groups give the BPQDs good dispersion stability in water.

In order to further analyze the phase structures and morphologies of different-sized BP-NS, transmission electron microscopy (TEM) was adopted (Figure 4). As shown in Figure 4, different-sized BP-NS, with sizes of about 2–4 μm (L-BP), 300–500 nm (M-BP), and 6–10 nm (BPQDs), were successfully prepared. Figure 4c,f,i reflects the HR-TEM images and SAED patterns of different-sized BP-NS. The lattice spacing of L-BP and M-BP is about 0.254 nm, corresponding to the (111) plane of BP. The lattice spacing of BPQDs is about 0.21 nm, corresponding to the (020) plane of BP. From the SAED patterns, it means that the different-sized BP-NS have well-defined diffraction rings, which clearly reflects the good crystallinity of the different-sized BP-NS.

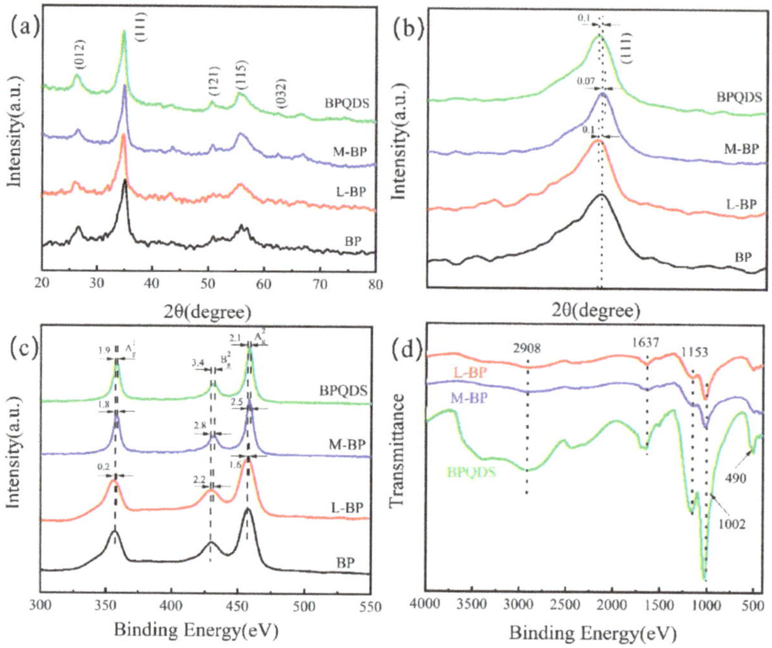

Figure 3. (**a**) XRD patterns, (**b**) the partial magnification of the (111) peak, (**c**) Raman spectrum, and (**d**) FTIR spectra of different-sized BP-NS.

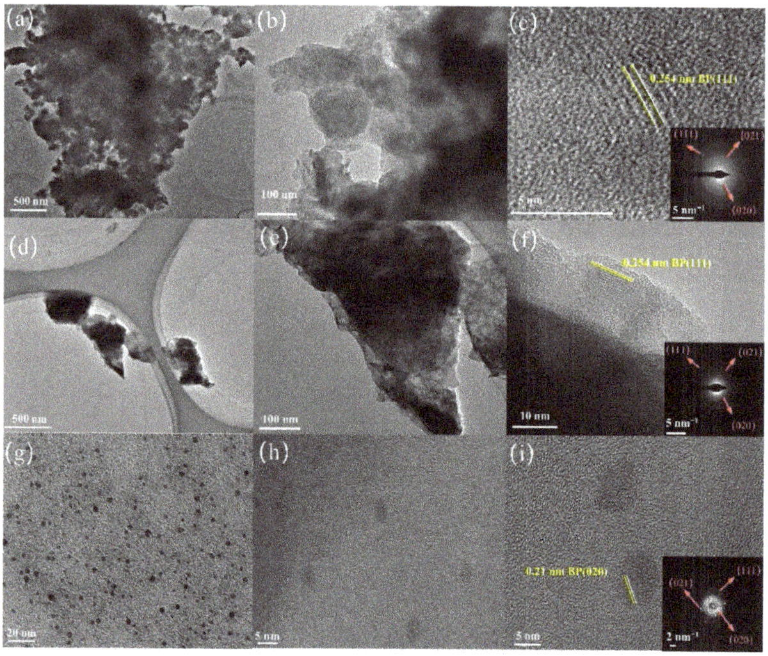

Figure 4. TEM images of the L-BP (**a–c**), M-BP (**d–f**), and BPQDs (**g–i**).

3.3. Tribological Performances

Figure 5a shows the COFs of ultrapure water and different-sized BP-NS as water-based lubrication additives. Compared with ultrapure water, the COFs of lubricants of different-sized BP-NS showed different degrees of reduction. Obviously, the COFs of 0.075 wt.% BPQD lubricants were more stable. The COFs of ultrapure water were firstly increased and then fluctuated over time. The average COF of ultrapure water is about 0.3252. At the time of 17 min, the COF of pure water was increased rapidly until the end of the frictional test. The reason is that during friction and wear, the water film was broken. It leads to the increase of the COF. After adding L-BP and M-BP into ultrapure water, the initial COFs were dropped rapidly and then remained at a stable value. When the small-sized BPQDs is added, in the initial stage, the COFs were lower and stable, indicating that the small-sized BPQDs could quickly enter the friction area and maintain stability. Figure 5b reflects the average COFs and wear rates of ultrapure water and lubricants of different-sized BP-NS. Obviously, L-BP, M-BP, and BPQDs are effective lubrication additives to reduce friction and wear. Compared with ultrapure water, adding 0.075 wt.% BPQDs lubrication additive could reduce COFs from 0.3252 to 0.1870, and the wear rates were decreased from 5.83×10^{-6} mm^3 to 1.03×10^{-6} mm^3 (Nm)$^{-1}$. The results show that the small-sized BPQDs present excellent wear resistance and can greatly decreased the COF and wear rate.

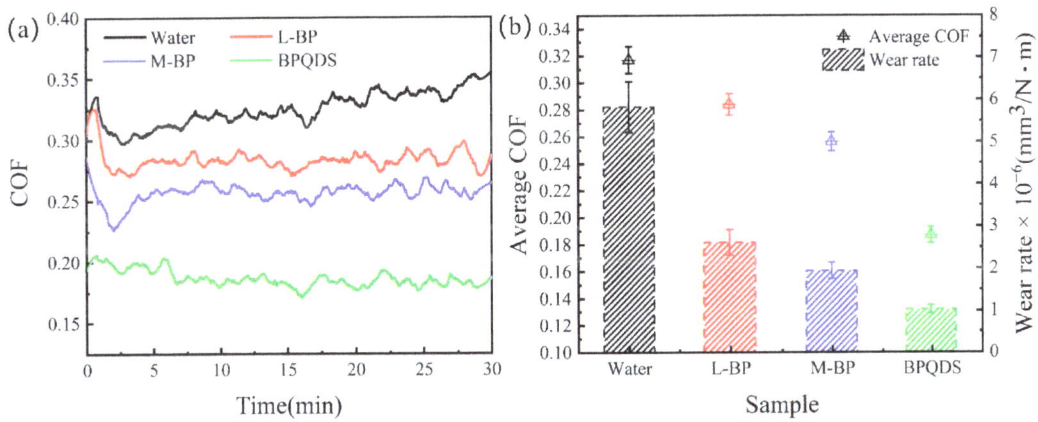

Figure 5. (**a**) COF versus sliding time. (**b**) Average COF and wear rate for the ultrapure water and different-sized BP-NS lubricants (0.075 wt.%, 10 N).

The concentration of the lubrication additives has great effect on their lubrication performance; therefore, the effects of the concentrations of the nano-additives on tribological properties were investigated. The COFs and wear rates of different concentrations of BPQD lubricants at the load of 10 N were reflected in Figure 6. As the BPQDs concentration increased, the changes of COF and wear rate show a trend of first decreasing and then increasing. As the concentration of BPQDs lubricating additive increased from 0.00 wt.% to 0.075 wt.%, the COF and wear rate decreased rapidly. The concentration of BPQDs lubricating additive increased from 0.075 to 0.1 wt.%, which may be due to the high concentration of BPQDs, which leads to particle aggregation, thus increasing the COFs and wear rates [27–30].

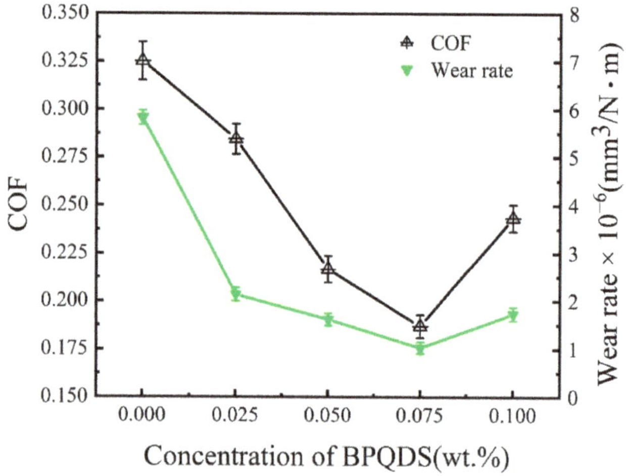

Figure 6. Average COFs and wear rates as a function of BPQDs concentration as water-based lubrication additives (10 N, 150 r/min, 0.5 h).

The tribological properties of different-sized BP-NS under different loads were investigated, as shown in Figure 7a. For ultrapure water, the COFs of L-BP and M-BP were decreased with increasing load. When the loads were increased from 10 N to 15 N, the average COFs of BPQDs increased slightly, although they were still lower than that of other lubrication additives. The reason is that because BPQDs lubrication additives achieve optimal tribological properties at 10 N, when the load increases, the wear rate and friction coefficient increase, which shows that BPQDs lubrication additives are in a boundary lubrication state at high load bearing, resulting in high friction and wear. Figure 7b shows the wear rates of different-sized BP-NS as water-based lubrication additives with different loads. It is not hard to see that the trend of change of wear rates closely resembled COFs'. Combined with the characteristics of COFs (Figure 5a) of different lubrication additives, it means that the short running-in time could lead to lower wear rates. In contrast, the wear rate is high due to the initial instability of the COFs during the friction process. The survey reflected that adding BP-NS into ultrapure water can reduce COF and wear rate. The BPQDs with small sizes have the best tribological properties.

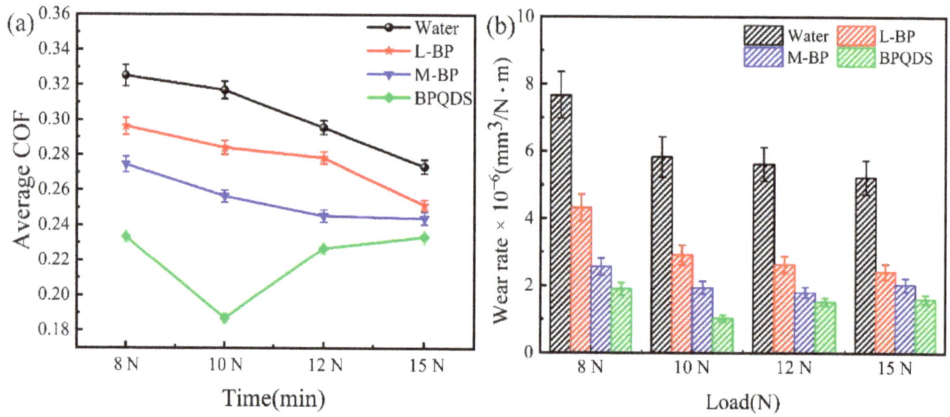

Figure 7. Tribological properties of different lubrication additives with different loads: (**a**) average COF, (**b**) wear rate (0.075 wt.%, 150 r/min, 0.5 h).

3.4. Worn Surface Analysis

The wear scar and wear spots lubricated by ultrapure water and lubricants of different-sized BP-NS are reflected in SEM (Figure 8). From the Figure 8a–d, it reflects that the ground spots are oval on the surface of the GCr15 ball. This can be explained that elliptical spots will be formed due to the differential hardness between the ball and the disc [31]. For ultrapure water and different-sized BP-NS, ultrapure water has the widest wear scar of 949 μm. Figure 8(a_2) shows that there are pits on the surface of the disc. After adding L-BP lubrication additives, the wear scar becomes shallower and the wear area becomes smaller. The serious furrow phenomenon was also happened. After adding M-BP lubrication additives, the width of the wear scar became narrower and the wear area became smaller, indicating that the M-BP exhibits better tribological properties at a load of 10 N. In sharp contrast, the scratches of the BPQD lubricants were observed to be the shallowest. Due to the small size effect, BPQDs are more likely to enter the friction contact area and undergo tribological chemical reactions, thereby reducing wear. The survey reflects that the smaller the nanosheets sizes can be obtained, the better tribological properties can be demonstrated.

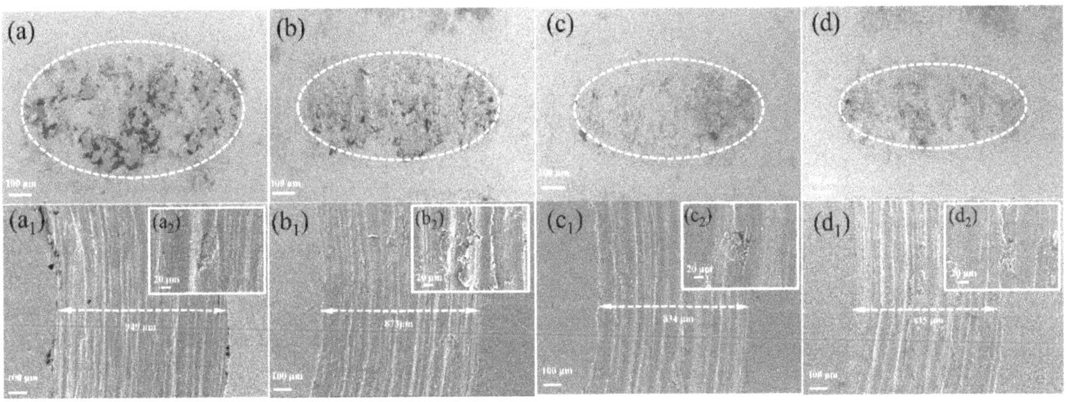

Figure 8. SEM images of worn surface of steel balls and the Ti6Al4V disc lubricated by ultrapure water (**a**,**a_1**,**a_2**), L-BP (**b**,**b_1**,**b_2**), M-BP (**c**,**c_1**,**c_2**), and BPQDs (**d**,**d_1**,**d_2**) (10 N, 150 rpm, 0.075 wt.%).

The EDS spectra of worn surface of Ti6Al4V disc lubricated by ultrapure water and lubricants of different-sized BP-NS are shown in Figure 9. The uniform distributions of Ti and Fe were presented on the wear scar obtained by ultrapure water. Besides these elements, the P element was found at the location of wear scars obtained by L-BP, M-BP, and BPQD lubricants, which means that the adsorption film was formed on the surface of the Ti6Al4V disc. It can be seen from the element content that when using BPQD lubricant, the P content on the surface of Ti6Al4V is up to 0.67 wt.%. This further illustrates that BPQDs can effectively enter the friction contact area to prevent direct contact between the upper and lower samples, thereby reducing friction and wear.

Figure 10 shows a 3D image depth map of the surface wear scars of a Ti6Al4V disc using different lubrication additives. As shown in Figure 10a, the maximum wear amount of ultrapure water is 0.467 mm^3, and the wear is relatively large, indicating that the lubrication ability of pure water is poor and corrosion wear occurs during friction. As can be seen from Figure 10b, when using L-BP lubricants, the depth of the wear scar is uneven, indicating the creation of deep grooves, resulting in an increase in wear. Using the M-BP lubricant in Figure 10c, the condition improved slightly, with flat wear scars and less wear. In contrast, using BPQD lubricants, the wear scars of the Ti6Al4V disc have the shallowest wear scars and the narrowest width, with a minimum wear volume of 0.233 mm^3 (Figure 10d).

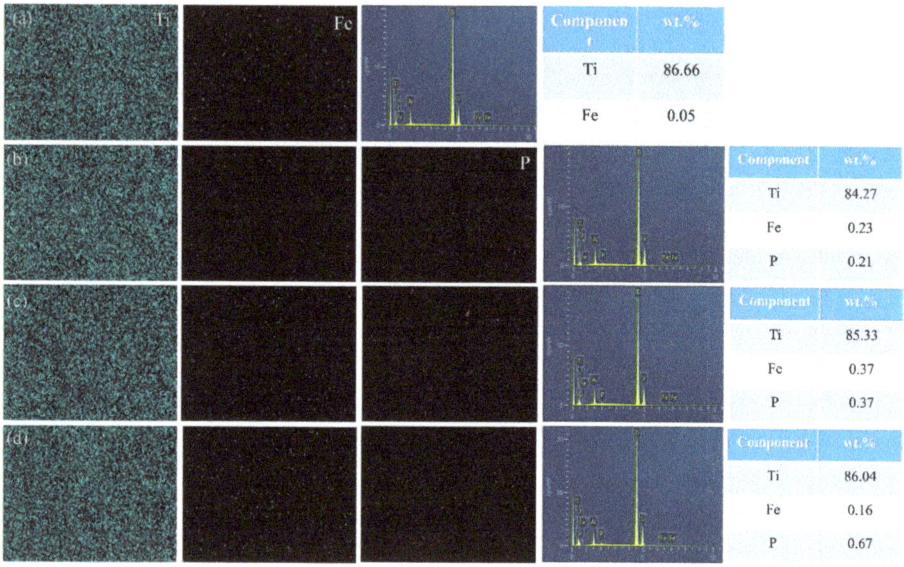

Figure 9. The EDS spectra of the worn surface of TI6AL4V discs lubricated by (**a**) Ultrapure Water, (**b**) L-BP, (**c**) M-BP, and (**d**) BPQD lubricant.

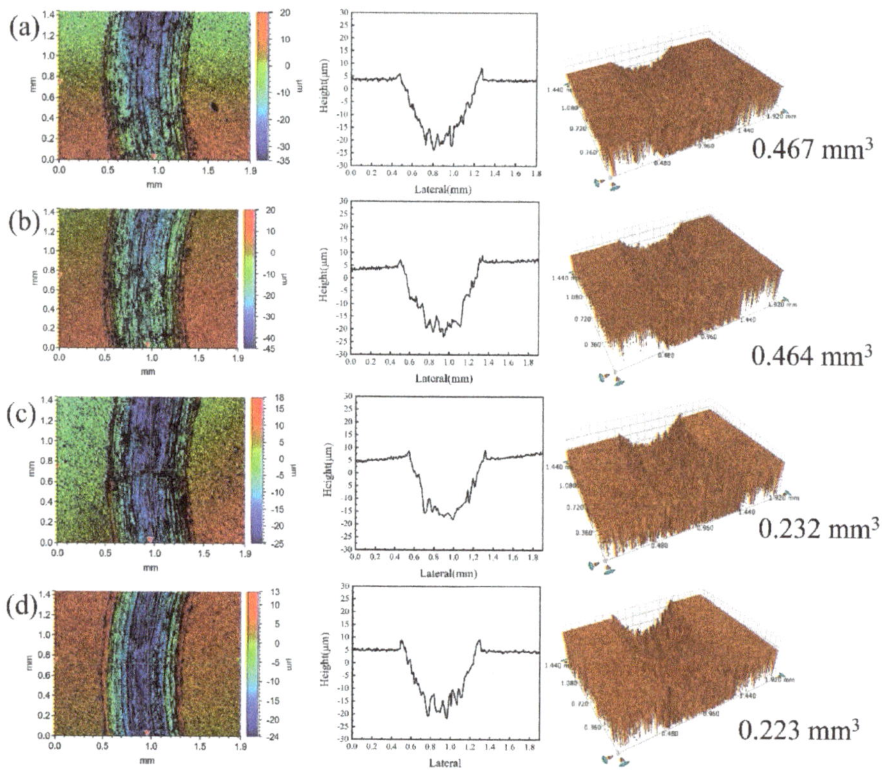

Figure 10. 3D depth maps of the wear scars on the discs lubricated by (**a**) ultrapure water (**b**) 075 wt.% L-BP, (**c**) 0.075 wt.% M-BP, and (**d**) 0.075 wt.% BPQDs (10 N, 150 r/min).

3.5. Lubrication Mechanism of the Lubrication Additives

In order to obtain the lubrication mechanisms of the different-sized BP-NS lubricants, Raman and XPS analysis was adopted. From the Raman spectra of Figure 11(a_1–c_1), the presence of L-BP, M-BP, and BPQDs has been confirmed by the characteristic peaks around 360 cm^{-1}, 430 cm^{-1}, and 460 cm^{-1} [32]. The survey indicates that an adsorption film is formed on the surface of Ti6Al4V, which has a protective effect on its surface. From Figure 11(a_3–c_3), it can indicate that the application of different lubricants produced TiO_2 and Al_2O_3, as can be perceived from the O1s peaks at 530.2 eV and 532.38 eV [33–35]. Additionally, from Figure 11(a_4–c_4), it can be seen that phosphorus oxides are produced on the friction surface [36]. However, when using BPQDs' lubrication additive, in addition to the above peaks, the GCr15 balls was oxidized to Fe_2O_3, which can be proved by the peak of 531.07 eV in the O1s spectrum in Figure 11(c_3) [37]. Additionally, in Figure 11(c_4), in addition to phosphorus oxides (133.38 eV), the 131.18 eV peak is derived from phosphate [26,36]. It can be inferred that the BPQDs lubrication additive have tribo-chemical reaction on the surface of Ti6Al4V disc and formed a tribo-chemical reaction film consisting of Fe_2O_3, Al_2O_3, TiO_2, phosphorus oxides, and phosphate on the wear surface.

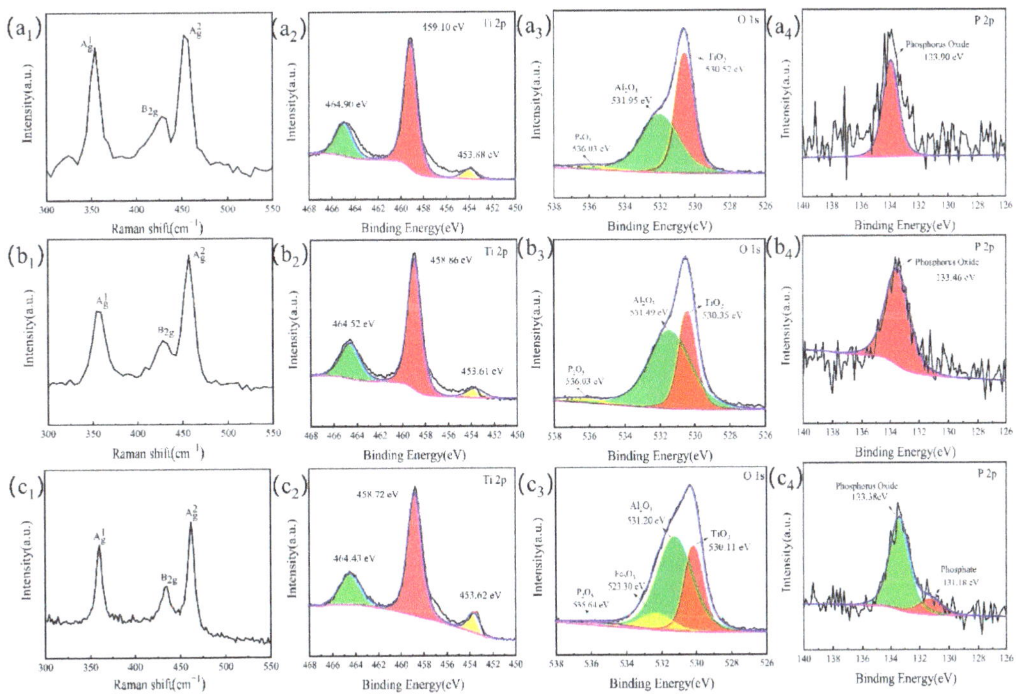

Figure 11. Raman and XPS of the wear scar surface lubricated by the 0.075 wt.% L-BP (**a_1–a_4**), M-BP (**b_1–b_4**), and BPQDs (**c_1–c_4**) lubrication additives (10 N, 150 r/min, 30 min).

The ratio of the film thickness (h_{min}) to the surface roughness (R_q) of the lubricated surface is used to determine the lubrication state of the frictional process, calculated as follows [38,39]:

$$\lambda = \frac{h_{min}}{R_q} \tag{4}$$

$$h_{min} = 2.8R'\left(\frac{\eta\mu_e}{E'R'}\right)^{0.65}\left(\frac{W_y}{E'R'^2}\right)^{-0.21} \tag{5}$$

$$R_q = \sqrt{R_{ball}^2 + R_{disc}^2} \quad (6)$$

where W_y and μ_e are the load (10 N) and sliding speed (64 mm/s), respectively, E' represents the modulus of elasticity (162 GPa), and R' represents the radius of ball (3 mm). Using the equation, the h_{min} is 7.6 nm. The calculated value of the λ is 0.166, which is less than 1, which explains that the lubrication condition is boundary lubrication. The tribological properties of the boundary lubrication are the interaction between the lubricant and the friction surface and the characteristic of the generating boundary lubrication film [40]. Therefore, the addition of lubricating additives into ultrapure water is an effective method to improve lubrication in metalworking.

By analyzing on the results of the surface of the wear scar, the lubrication mechanisms of the different-sized of BP-NS lubricants are proposed (Figure 12). According to XPS and Raman results, the tribological layers including the adsorption film of black phosphorus and the tribo-chemical film were produced on the surface of a Ti6Al4V titanium alloy during the friction process. The lubrication mechanism of L-BP and M-BP is due to the shearing effect between the BP nanosheets to prevent the direct contact of the frictional pairs. The good lubricity of BPQDs is due to its small size for easily accessing to the contact area. The chemical reaction films consisting of Fe_2O_3, Al_2O_3, TiO_2, phosphorus oxides, and phosphates are beneficial to improving lubrication properties. The formed tribological layers can greatly prevent the contact of the friction surface [41–43]. In addition, BPQDs can also be used as the ball bearing in the friction process to decrease friction and wear [44,45]. In conclusion, the reason for the excellent friction performance of BPQD lubricant is that the small size of BPQDs is more likely to enter the contact area to produce an effective tribo-chemical film on the surface of the friction pairs.

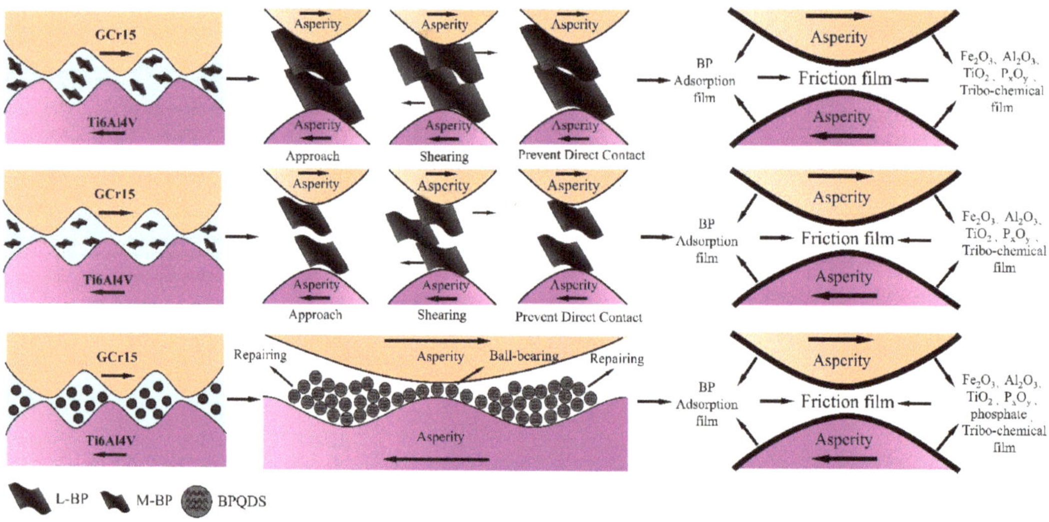

Figure 12. Schematic of lubrication mechanism of L-BP, M-BP, and BPQD lubricants.

4. Conclusions

In general, the different-sized of BP-NS are prepared by HEBM and liquid phase exfoliation methods. The tribological properties of the different-sized of BP-NS as water-based lubrication additives are evaluated under different loads for GCr15/Ti6Al4V contact. The results show that the COFs and wear rates of the different-sized of BP-NS lubricants are decreased to different degrees. The 0.075 wt.% BPQD lubricant exhibits the best tribological properties. Compared with ultrapure water, the COF and wear rate are reduced by about 42.4% and 82.3%, respectively. The lubrication mechanisms can ascribe to the BP adsorption film and the tribo-chemical film. On the one hand, BPQDs are more likely to enter the contact area. On the other hand, the anti-wear performance of Ti6Al4V depends on the adsorption film and tribo-chemical reaction film. These aspects prevent direct contact on the friction pairs, and effectively improve anti-friction and anti-wear, showing the excellent tribological performance of BPQDs.

Author Contributions: Methodology, W.W.; validation, S.D. and W.W.; investigation, Y.G.; writing—original draft preparation, S.D.; writing—review and editing, S.D., W.W. and G.D. All authors have read and agreed to the published version of the manuscript.

Funding: This work was supported by the fund of the Youth Science and Technology New Star Project of Shaanxi Province Innovation Ability Support Plan (2021KJXX-32) and Advanced technology research program of Xi'an (21XJZZ0031).

Institutional Review Board Statement: Not applicable.

Informed Consent Statement: Written informed consent has been obtained from the patient(s) to publish this paper.

Data Availability Statement: All the original data and graphic files of this study are stored in Figshare. And the DOI number is [10.6084/m9.figshare.19102403].

Conflicts of Interest: The authors declare no conflict of interest.

References

1. Chong, Y.; Deng, G.; Gao, S.; Yi, J.; Shibata, A.; Tsuji, N. Yielding nature and Hall-Petch relationships in Ti-6Al-4V alloy with fully equiaxed and bimodal microstructures. *Scr. Mater.* **2019**, *172*, 77–82. [CrossRef]
2. Deng, G.; Zhao, X.; Su, L.; Wei, P.; Zhang, L.; Zhan, L.; Chong, Y.; Zhu, H.; Tsuji, N. Effect of high pressure torsion process on the microhardness, microstructure and tribological property of Ti6Al4V alloy. *J. Mater. Sci.* **2021**, *94*, 183–195. [CrossRef]
3. Souza, P.M.; Mendiguren, J.; Chao, Q.; Beladi, H.; Hodgson, P.D.; Rolfe, B.A. Microstructural based constitutive approach for simulating hot deformation of Ti6Al4V alloy in the $\alpha + \beta$ phase region. *Mater. Sci. Eng. A* **2019**, *748*, 30–37. [CrossRef]
4. Hong, J.I.A.; Fu-sheng, L.U.; Bin, H.A.O. Report on China Titanium industry progress in 2020. *Titianium Ind. Prog.* **2021**, *38*, 34–41.
5. Koizumi, H.; Takeuchi, Y.; Imai, H.; Kawai, T.; Yoneyama, T. Application of titanium and titanium alloys to fixed dental prostheses. *J. Prosthodont. Res.* **2019**, *63*, 266–270. [CrossRef] [PubMed]
6. Davim, J.P. *Machining of Titanium Alloys*; Springer: Berlin, Germany, 2014.
7. Singh, P.; Pungotra, H.; Kalsi, N.S. On the characteristics of titanium alloys for the aircraft applications—ScienceDirect. *Mater. Today Proc.* **2017**, *4*, 8971–8982. [CrossRef]
8. Yang, Y.; Zhang, C.; Dai, Y.; Luo, J. Tribological properties of titanium alloys under lubrication of SEE oil and aqueous solutions. *Tribol. Int.* **2017**, *109*, 40–47. [CrossRef]
9. Cheng, J.; Yang, J.; Ma, J.; Bi, Q.; Zhang, X.; Fu, L.; Li, F.; Zhu, S.; Liu, W. The tribological behavior of a Ti-46Al-2Cr-2Nb alloy under liquid paraffine lubrication. *Tribol. Lett.* **2012**, *46*, 233–241. [CrossRef]
10. Xie, H.; Wei, Y.; Jiang, B.; Tang, C.; Nie, C. Tribological properties of carbon nanotube/SiO2 combinations as water-based lubricant additives for magnesium alloy. *J. Mater. Res. Technol.* **2021**, *12*, 138–149. [CrossRef]
11. Sharma, M.D.; Sehgal, R. Experimental study of friction and wear characteristics of titanium alloy (Ti-6Al-4V) under lubricated sliding condition. *Ind. Lubr. Tribol.* **2014**, *66*, 174–183. [CrossRef]
12. Wang, Q.; Hou, T.; Wang, W.; Zhang, G.; Gao, Y.; Wang, K. Tribological behavior of black phosphorus nanosheets as water-based lubrication additives. *Friction* **2022**, *10*, 374–387. [CrossRef]
13. Wang, W.; Dong, S.; Gao, Y.; Zhang, G.; Wang, K. Tribological behaviours of black phosphorus/MoS_2 composites as water-based lubrication additives. *Lubr. Sci.* **2021**, *3*, 404–416. [CrossRef]
14. Li, Z.; Ma, S.; Zhang, G.; Wang, D.; Zhou, F. Soft/Hard-Coupled Amphiphilic Polymer Nanospheres for Water Lubrication. *ACS Appl. Mater. Interfaces* **2018**, *10*, 9178–9187. [CrossRef] [PubMed]
15. Podgornik, B.; Kafexhiu, F.; Kosec, T.; Jerina, J.; Kalin, M. Friction and anti-galling properties of hexagonal boron nitride (h-BN) in aluminium forming. *Wear* **2017**, *388*, 2–8. [CrossRef]
16. Xie, G.; Luo, J.; Guo, D.; Liu, S. Nanoconfined ionic liquids under electric fields. *Appl. Phys. Lett.* **2010**, *96*, 2071. [CrossRef]
17. Xu, Y.; Nan, F.; Xu, B.S. Tribological properties of attapulgite/oil-soluble nano-Cu composite lubrication additive. *J. Mater. Eng.* **2016**, *44*, 41–46.
18. Xie, H.; Jiang, B.; Liu, B.; Wang, Q.; Xu, J.; Pan, F. An investigation on the tribological performances of the SiO_2/MoS_2 hybrid nanofluids for magnesium alloy-steel contacts. *Nanoscale Res. Lett.* **2016**, *11*, 329–346. [CrossRef]
19. Pu, J.B.; Wang, L.P.; Xue, Q.J. Progress of tribology of graphene and graphene-based composite lubrication materials. *Tribology* **2014**, *34*, 93–112.

20. Xing, C.; Zhang, J.; Jing, J.; Li, J.; Shi, F. Preparations, properties and applications of low-dimensional black phosphorus. *Chem. Eng. J.* **2019**, *370*, 120–135. [CrossRef]
21. Qu, G.; Liu, W.; Zhao, Y.; Gao, J.; Xia, T.; Shi, J.; Hu, L.; Zhou, W.; Gao, J.; Wang, H.; et al. Improved biocompatibility of black phosphorus nanosheets by chemical modification. *Angew. Chem. Int. Ed.* **2017**, *56*, 14488–14493. [CrossRef]
22. Wang, W.; Xie, G.; Luo, J. Black phosphorus as a new lubricant. *Friction* **2018**, *6*, 116–142. [CrossRef]
23. Wang, W.; Xie, G.; Luo, J. Superlubricity of Black Phosphorus as Lubricant Additive. *ACS Appl. Mater. Interfaces* **2018**, *10*, 43203–43210. [CrossRef] [PubMed]
24. Wu, S.; He, F.; Xie, G.; Bian, Z.; Luo, J.; Wen, S. Black phosphorus: Degradation favors lubrication. *Nano Lett.* **2018**, *18*, 5618–5627. [CrossRef] [PubMed]
25. Ren, X.; Yang, X.; Xie, G.; Luo, J. Black phosphorus quantum dots in aqueous ethylene glycol for macroscale superlubricity. *ACS Appl. Nano Mater.* **2020**, *3*, 4799–4809. [CrossRef]
26. Tang, W.; Jiang, Z.; Wang, B.; Li, Y. Black phosphorus quantum dots: A new-type of water-based high-efficiency lubricant additive. *Friction* **2020**, *9*, 1528–1542. [CrossRef]
27. Wu, H.; Zhao, J.; Cheng, X.; Xia, W.; He, A.; Yun, J.H.; Huang, S.; Wang, L.; Huang, H.; Jiao, S. Friction and wear characteristics of TiO_2 nano-additive water-based lubricant on ferritic stainless steel. *Tribol. Int.* **2018**, *117*, 24–38. [CrossRef]
28. Peng, D.X.; Chen, C.H.; Kang, Y.; Chang, Y.P.; Chang, S.Y. Size effects of SiO_2 nanoparticles as oil additives on tribology of lubricant. *Ind. Lubr. Tribol.* **2013**, *62*, 111–120. [CrossRef]
29. Peng, D.X.; Chen, C.H.; Kang, Y.; Chang, Y.P.; Chang, S.Y. Tribological studies of transmission oil dispersed with molybdenum disulfide and tungsten disulfide nanoparticles. *J. Tribol.* **2017**, *139*, 041301.
30. Xie, H.; Jiang, B.; He, J.; Xia, X.; Pan, F. Lubrication performance of MoS_2 and SiO_2 nanoparticles as lubricant additives in magnesium alloy-steel contacts. *Tribol. Int.* **2016**, *93*, 63–70. [CrossRef]
31. Wang, W.; Zhang, G.; Xie, G. Ultralow concentration of graphene oxide nanosheets as oil-based lubricant additives. *Appl. Surf. Sci.* **2019**, *498*, 143683. [CrossRef]
32. Rissi, E.N.; Soignard, E.; McKiernan, K.A.; Benmore, C.J.; Yarger, J.L. Pressureinduced crystallization of amorphous red phosphorus. *Solid State Commun.* **2012**, *152*, 390–394. [CrossRef]
33. Sampath, S.; Maydannik, P.; Ivanova, T.; Homola, T.; Sillanpää, M.; Nagumothu, R.; Alagan, V. Structural and morphological characterization of Al2O3 coated macro-porous silicon by atomic layer deposition. *Thin Solid Film.* **2016**, *616*, 628–634. [CrossRef]
34. Boonprakob, N.; Wetchakun, N.; Phanichphant, S.; Waxler, D.; Sherrell, P.; Nattestad, A.; Chen, J.; Inceesungvorn, B. Enhanced Visible-light Photocatalytic Activity of g-C_3N_4/TiO_2 Films. *J. Colloid Interface Sci.* **2014**, *417*, 402. [CrossRef]
35. Xie, H.; Dang, S.; Jiang, B.; Xiang, L.; Zhou, S.; Sheng, H.; Yang, T.; Pan, F. Tribological performances of SiO2/graphene combinations as water-based lubricant additives for magnesium alloy rolling. *Appl. Surf. Sci.* **2019**, *475*, 847–856. [CrossRef]
36. Ni, H.; Liu, X.; Cheng, Q.D. A New Strategy for Air-Stable Black Phosphorus Reinforced PVA Nanocomposites. *J. Mater. Chem. A* **2018**, *6*, 7142–7147. [CrossRef]
37. Meng, Y.; Su, F.; Chen, Y. A novel nanomaterial of graphene oxide dotted with Ni nanoparticles produced by supercritical CO_2-assisted deposition for reducing friction and wear. *ACS Appl. Mater. Interfaces* **2015**, *7*, 11604–11612. [CrossRef] [PubMed]
38. Gong, K.; Wu, X.; Zhao, G.; Wang, X. Nanosized MoS_2 deposited on graphene as lubricant additive in polyalkylene glycol for steel/steel contact at elevated temperature. *Tribol. Int.* **2017**, *110*, 1–7.
39. Guo, D.; Li, J.; Xie, G.; Wang, Y.; Luo, J. Elastic properties of polystyrene nanospheres evaluated with atomic force microscopy: Size effect and error analysis. *Langmuir ACS J. Surf. Colloids* **2014**, *30*, 7206–7212. [CrossRef]
40. Si, L.; Guo, D.; Luo, J.; Lu, X. Monoatomic layer removal mechanism in chemical mechanical polishing process: A molecular dynamics study. *J. Appl. Phys.* **2010**, *107*, 3115. [CrossRef]
41. Hu, Y.; Wang, Y.; Wang, C.; Ye, Y.; Zhao, H.; Li, J.; Lu, X.; Mao, C.; Chen, S.; Mao, J.; et al. One-pot pyrolysis preparation of carbon dots as eco-friendly nanoadditives of water-based lubricants. *Carbon* **2019**, *152*, 511–520. [CrossRef]
42. Zhou, Y.; Qu, J. Ionic liquids as lubricant additives: A review. *ACS Appl. Mater. Interfaces* **2017**, *9*, 3209–3222. [CrossRef] [PubMed]
43. Tu, Z.; Hu, E.; Wang, B.; David, K.D.; Seeger, P.; Moneke, M.; Stengler, R.; Hu, K.; Hu, X. Tribological behaviors of Ni-modified citric acid carbon quantum dot particles as a green additive in polyethylene glycol. *Friction* **2020**, *8*, 182–197. [CrossRef]
44. Dai, W.; Kheireddin, B.; Gao, H.; Liang, H. Roles of nanoparticles in oil lubrication. *Tribol. Int.* **2016**, *102*, 88–98. [CrossRef]
45. He, C.; Yan, H.; Li, X.; Wang, X. In situ fabrication of carbon dots-based lubricants using a facile ultrasonic approach. *Green Chem.* **2019**, *21*, 2279–2285. [CrossRef]

metals

Article

Formulation of Sustainable Water-Based Cutting Fluids with Polyol Esters for Machining Titanium Alloys

Elisabet Benedicto [1,2,*,†], Eva María Rubio [1], Laurent Aubouy [2] and María Ana Sáenz-Nuño [3]

1 Department of Manufacturing Engineering, Industrial Engineering School, Universidad Nacional de Educación a Distancia (UNED), St/Juan del Rosal 12, E28040 Madrid, Spain; erubio@ind.uned.es
2 Department of Surface Chemistry, Leitat Technological Center, C/Innovació 2, E08225 Terrassa, Spain; laubouy@leitat.org
3 Department of Manufacturing Engineering, Instituto de Investigación Tecnológica, IIT (Comillas Pontifical University), St/Alberto Aguilera 25, E28015 Madrid, Spain; msaenz@comillas.edu
* Correspondence: ebenedict6@alumno.uned.es; Tel.: +34-660-127-618
† Programa de Doctorado en Tecnologías Industriales.

Abstract: The machinability of titanium alloys still represents a demanding challenge and the development of new clean technologies to lubricate and cool is greatly needed. As a sustainable alternative to mineral oil, esters have shown excellent performance during machining. Herein, the aim of this work is to investigate the influence of esters' molecular structure in oil-in-water emulsions and their interaction with the surface to form a lubricating film, thus improving the efficiency of the cutting fluid. The lubricity performance and tool wear protection are studied through film formation analysis and the tapping process on Ti6Al4V. The results show that the lubricity performance is improved by increasing the formation of the organic film on the metal surface, which depends on the ester's molecular structure and its ability to adsorb on the surface against other surface-active compounds. Among the cutting fluids, noteworthy results are obtained using trimethylolpropane trioleate, which increases the lubricating film formation (containing 62% ester), thus improving the lubricity by up to 12% and reducing the torque increase due to tool wear by 26.8%. This work could be very useful for fields where often use difficult-to-machine materials—such as Ti6Al4V or γ-TiAl—which require large amounts of cutting fluids, since the formulation developed will allow the processes to be more efficient and sustainable.

Keywords: cutting fluid; esters; lubrication; tool wear; titanium alloys

Citation: Benedicto, E.; Rubio, E.M.; Aubouy, L.; Sáenz-Nuño, M.A. Formulation of Sustainable Water-Based Cutting Fluids with Polyol Esters for Machining Titanium Alloys. *Metals* **2021**, *11*, 773. https://doi.org/10.3390/met11050773

Academic Editor: Guanyu Deng

Received: 16 April 2021
Accepted: 5 May 2021
Published: 8 May 2021

1. Introduction

Titanium alloys, particularly Ti6Al4V, are used extensively in the aeronautic, aerospace, automotive, chemical, and biomedical industries due to their unique mechanical properties (Table 1), such as high specific strength, hardness, corrosion resistance at high temperatures, and biocompatibility [1]. Despite these exceptional properties, Ti6Al4V is classified as a difficult-to-cut material because of its high chemical affinity, low thermal conductivity, and work hardening. The poor machinability of these alloys leads to excessive tool wear and catastrophic tool failure, resulting in decreased tool life and low productivity.

Due to the low conductivity of Ti6Al4V, the heat generated during machining cannot be dissipated throughout the workpiece and chips effectively. Therefore, the application of lubrication/cooling systems is extremely important. In most difficult-to-machine alloys, cutting fluids are employed in the machining process by providing lubrication under boundary friction at the workpiece-tool interface, or, more specifically, at the chip-tool interface, eliminating heat from the cutting zone and removing abrasive particles from the contact area [2]. The creation of the lubricating film depends on the media composition, particularly the absorption and chemisorption of polar or metal-active additives on the surface of the workpiece, while the coolant and cleaning capacity depends on the physical properties of the cutting fluids (e.g., specific heat capacity, vaporization heat, viscosity, and surface tension) [3]. Additionally, the cutting fluid protects the workpiece, the tool, the machine tool, and other metal surfaces from corrosion.

Table 1. The main properties of the Ti6Al4V alloy [4–6].

Property	Ti6Al4V
Density (g/cm^3)	4.42
Young's modulus at RT (GPa)	120 ± 10
Yield Strength (MPa)	800–1100
Tensile strength (MPa)	900–1200
Ductility at RT (%)	13–16
Creep limit (°C)	385
Oxidation limit (°C)	400
Hardness (HV10)	360 ± 30
Thermal conductivity (W·mK)	7.5–8.6

However, the use of conventional mineral oil-based cutting fluids has been thoroughly reviewed because of their environmental and health hazards [7]. Many investigations have aimed to improve the machinability of titanium alloys using sustainable lubrication/cooling systems. The advantages and drawbacks of the most common systems are summarized in Table 2.

Table 2. The advantages and drawbacks of sustainable lubrication/cooling systems used in titanium alloy machining [1,8,9].

Lubrication/ Cooling Systems	Advantages	Drawbacks
Dry cutting	No need for cutting fluid Easier chip collection for recycling Minimal environmental impact	High cutting temperature generation High tendency of workpiece microstructural alterations Reliable results for limited cutting parameters Poor tool life and surface finish Problematic chip evacuation
MQL	Reduction in cutting fluid Less costly method in comparison to other systems Good results in terms of cutting forces, tool wear, surface roughness are noted	Poor chip evacuation Poor cooling capacity Mist formation Very sensitive to MQL supply system
Cryogenic cooling	Eliminate the use of cutting fluid No need to clean the chips and improved chip breakability Promote improvements in surface integrity Improved tool life	Highly sensitive to tool–material pairs. The production cost of the cryogen is very high compared to cutting fluids Special Dewar is needed for cryogenic supply Overcooling lead to embrittlement of workpiece
Cold compressed air	Absence of cutting fluid Chips can be collected in dry form	Additional energy is required to produce compressed air
Sustainable cutting fluids	Can be totally biodegradable and renewable Less costly compared to cryogenic cooling Chip evacuation Effective removal of heat	Vegetable oils have low oxidation stability Formulation difficulties

A great improvement in the machining of titanium alloys is observed when different lubrication/cooling systems are combined. Noteworthy results were obtained with a minimum-quantity lubrication (MQL) and a cryogenic cooling hybrid system. The poor cooling capacity and lack of chip evacuation in the MQL system can be enhanced using compressed air or cryogenic cooling [10]. Moreover, vegetable oils or even recycled oils can be used as MQL fluid, thus increasing not only its efficiency but also reducing its environmental impact [11].

In order to ensure high levels of productivity whilst meeting the surface integrity of the machined parts, industrial companies still employ cutting fluids, especially in the most demanding operations. More than 2,000,000 m^3 of cutting fluids are used globally each year, the majority of which (85%) are petroleum-based [12]. Hence, these cutting fluids are non-renewable and toxic. Therefore, introducing new sustainable materials for the formulation of environmentally friendly cutting fluids as an alternative to mineral oil-based fluids is one of the main future trends in the machining of titanium alloys [10].

When selecting the type of cutting fluid to use, there are several aspects to consider, such as the machining operation, the workpiece material, the cutting tool material, the mode of application, the cutting fluid, and the environment friendliness [13]. The excellent heat dissipating properties of water and its cleaning capacity to remove chips of water-based cutting fluid, make it suitable for machining titanium alloys [14]. Moreover, the components of the cutting fluid and its physico-chemical properties determine the wetting and adsorption properties on the metal surface, which further affect the quality and performance of the machined surface. Chemical reactions can result in the loss of effective alloying elements thus damaging the surface [15].

Sustainable cutting fluids is a commonly used term for products that meet the following characteristics: good biodegradability; low eco-toxicity; risk of low contamination for water, air. and soil; low consumption; a long shelf life; recyclable; an ability to produce less waste; and an ability to promote energy saving [16]. There is a trend towards the use of vegetable or synthetic based oils and against the formulation of environmentally hazardous mineral oils [17], as well as the elimination or reduction in hazardous substances in the formulation of cutting fluids.

The most commonly used sustainable base fluids as alternatives to mineral bases are low molecular weight polyalphaolefins, polyalkylene glycols, vegetable oils, and synthetic esters [18]. This research addresses synthetic esters, which have attracted wider interest from both academic researchers and industrial users due to their high polarity and excellent lubricity in the boundary lubrication [19]. Synthetic esters perform well over a large temperature range, have a high viscosity index, possess a high lubricity, provide corrosion protection, and have a high oxidative stability [20]. In general, synthetic esters meet the requirements for aquatic biodegradability and toxicity, although they tend to be less readily biodegradable than vegetable oil-based lubricants.

There is a wide range of synthetic esters with properties that vary greatly depending on their chemical structure. A synthetic ester in its simplest form consists of an alcohol and a fatty acid. In lubricants, esters are usually made with two or more carboxylate groups. Due to the large number of different alcohols and fatty acids available for ester formulation, esters can be synthesized to suit a specific application [21]. In particular, this study is focused on polyol esters obtained from the reaction of fatty acids and polyhydric alcohols, also known as polyol, which are less susceptible to hydrolysis [22]. The fatty acids most commonly considered in the synthesis of polyol esters are caprylic acid, oleic acid, rapeseed oil, olive oil, animal fats, and palm oil. Examples of commonly used polyols are trimethylolpropane, neopentylglycol, and pentaerythritol [23].

The challenges in formulating water-based cutting fluids continue to increase as end users demand a better performance over longer periods of time and under harsh conditions [24]. Currently, vegetable oils and the synthetic esters obtained from them are emulsified as additives in water-based cutting fluids [25]. However, there are few studies that describe the lubrication/cooling results obtained in the machining of difficult-to-cut materials [26].

In contrast to oils, water has many unique properties due to its polarity, which makes aqueous lubrication much more complicated than traditional oil lubrication. Water-based lubrication mechanisms are not yet well known due to the diversity and complexity of aqueous solutions. In addition, some traditional lubrication theories are not applicable to aqueous lubrication due to the complex physical and chemical properties of water [27]. In the case of water-soluble cutting fluids, the effect of hydrolysis must be considered. Emulsions contain significant amounts of water, which can cause ester hydrolysis and variation of the chemical composition in the lubrication film, resulting in a completely different tribochemical mechanism between the water-based emulsion and the ester [28].

The aim of this work was to study the influence of the polyol ester on the development of a sustainable cutting fluid for titanium alloys. Oil-in-water (O/W) emulsions were formulated with several polyol esters and the ability to form a lubrication film was analyzed on Ti6Al4V strips. Finally, the lubricity performance was tested using a tapping process and evaluated with design of experiments (DOE) software.

2. Materials and Methods

2.1. Materials and Sample Preparation

To carry out the study, five cutting fluids were prepared by adding a mixture of non-ionic and anionic surfactants to stabilize the O/W emulsions. Oleyl/cetyl propoxylated alcohol (BASF, Ludwigshafen, Germany) and oleth-10 carboxylic acid (Kao Chemicals GmbH, Emmerich, Germany) were mixed in deionized water at 0.8 mmol/L and 1.2 mmol/L, respectively. The mixture was adjusted to the recommended pH 9.2 with monoethanolamine.

Four commercial esters synthetized by Industrial Química Lasem (IQL, Castellgalí, Spain) were used to formulate the cutting fluids. All of them were commercialized as environmentally adapted lubricants [29] and were recommended for the formulation of lubricants. The esters were obtained from a reaction of oleic acid and the number of ramifications were varied, modifying the alcohol group. The esters were added in a concentration of 1.0 mmol/L and stirred in order to obtain a homogeneous mixture. The same molar concentration for all esters was used to easily compare the effect of the molecular structure of the polyol esters. The cutting fluids prepared were:

- without an ester;
- C18:1 × 1 using isopropyl oleate;
- C18:1 × 2 using neopentylglycol dioleate;
- C18:1 × 3 using trimethylolpropane trioleate;
- C18:1 × 4 using pentaerythrityl tetraoleate.

Moreover, the cutting fluids were prepared without the addition of anti-wear and extreme pressure chemical additives to prevent interference with the lubricity performance of the mixtures.

2.2. Determination of Lubricating Film on Ti6Al4V Surface

The lubricity performance is related to the capability to form a lubricating film on the surface, providing a layer that protects the surface from wear. The schematic diagram of the experimental setup is shown in Figure 1. Tests were conducted with Ti6Al4V (grade 5) panels (special metals and products, Spain) with dimensions of 17 × 75 × 0.8 mm^3, and the chemical composition shown in Table 3. Panels were immersed in the cutting fluid at room temperature and stirred for 10 min. The film formation was determined according to the method developed by Benedicto et al. [30]: (1) measuring the total organic carbon (*TOC*) with RC612 (Leco, St. Joseph, MI, USA) to quantify the milligrams of carbon adhered to the Ti6Al4V panel and (2) calculating the ratio of the ester, analyzed by infrared reflection absorption spectroscopy (IRRAS) (Vertex 70 Bruker, Ettlingen, Germany).

Table 3. The chemical composition of Ti6Al4V (grade 5) (wt%).

Ti	Al	V	Fe	O
89.75%	6%	4%	0.25% max.	0.2% max

Following this method, four calibration curves were built for each ester, varying its concentration in the O/W emulsion and using a Fourier Transform infrared spectrophotometer (Iraffinity-1S Shimadzu, Nagoya, Japan). The regression line provided the ratio (*REO*) between the integrated absorbance under the C=O peak (1735–1750 cm^{-1}) and under the CH_2 stretching vibration peaks (2850–2925 cm^{-1}). The following regression equation allowed the percentage of ester (wt%) to be given a *REO* value from a spectrum and the correlation coefficient (R^2) measured the strength of the relationship between them:

$$\% \text{ ester} = A + B \cdot REO, \quad (1)$$

Finally, the amount of ester adhered to the Ti6Al4V panels was calculated by multiplying the percentage of ester by the milligrams of *TOC*, taking into account the molecular weight and the average of number of carbons on the organic layer for each cutting fluid.

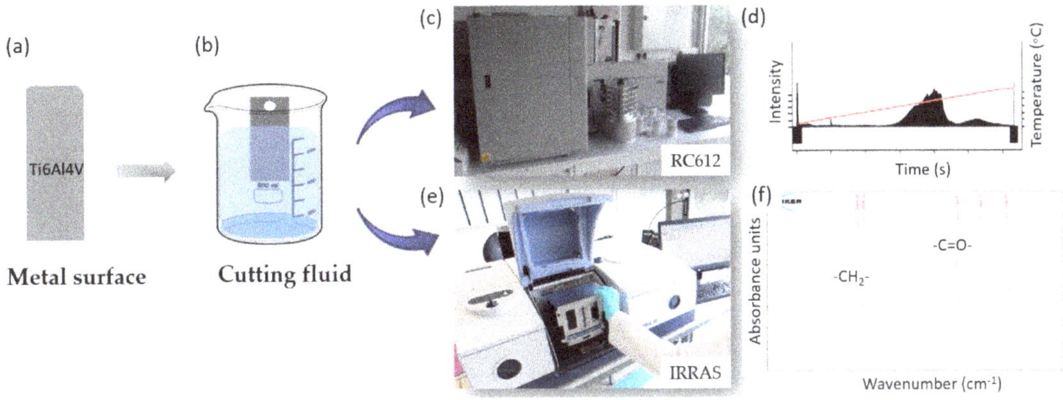

Figure 1. The experimental setup to study the lubrication film formed on the panel surface: (**a**)Ti6Al4V panel, (**b**) cutting fluid, (**c**) RC612 equipment, (**d**) the total carbon rate of evolution as a function of temperature, (**e**) IRRAS equipment and, (**f**) IRRAS spectra with C=O and CH_2 stretching vibration peaks.

2.3. *Lubricity Performance of Polyol Ester Cutting Fluids*

To gain a better understanding of the behavior of the polyol ester in water-based cutting fluids, a tapping torque test was conducted in Labtap G8 (Microtap, Munich, Germany). This cutting operation is highly sensitive to lubrication [31]. The material used for the machining was a Ti6Al4V with pre-drilled holes (Figure 2). A TTT_M4C coated tool, size M4 (0.7 mm pitch, 3.3 mm tapping diameter) with helical channels was used for each cutting fluid.

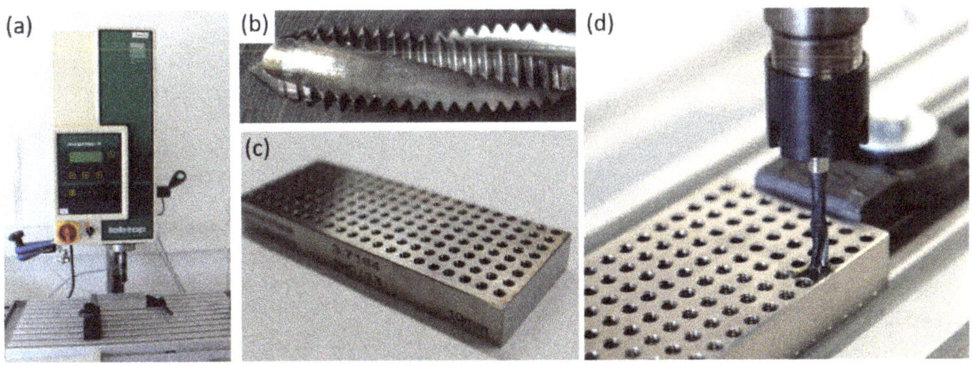

Figure 2. The experimental setup to study the lubricity performance using the Labtap G8: (**a**) Labtap G8 equipment, (**b**) TTT_M4C coated tool, (**c**) Ti6Al4V workpiece and, (**d**) cutting process.

The cutting fluid was poured in the holes to lubricate them during the tapping process at 300 rpm, with a 6 mm length of thread, as shown in Table 4 each tapping process was repeated 15 times or until the tool broke. Figure 3 shows the tapping procedure graphically where: (1) shows the beginning of the cut; (2) shows the tool penetrating the workpiece and the torque increasing due to the increasing contact surface between the workpiece and the tool; (3) shows the tool cutting with all its chamber teeth until the length's thread is achieved and (4) shows the beginning of the reversal of the spindle to bring the tool to the initial position [32]. Finally, the results were reported averaging the tapping torque values (N·cm) in the 1 mm to 6 mm range of cut.

Table 4. Cutting operation conditions.

Parameter	Value
Workpiece material	Ti6Al4V (grade 5) pre-drilled
Spindle speed (rpm)	300
Length of the thread (mm)	6
Hole diameter (mm)	3.3
Tapping tool	TTT_M4C coated tool. Size M4
Tap mode	Cutting

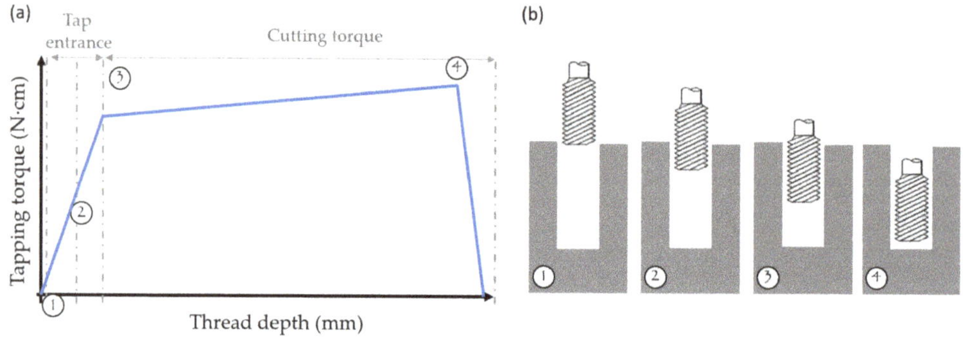

Figure 3. The tapping process: (**a**) graphical measurement of the tapping torque, (**b**) different positions of the tap during processing. (1) beginning of the cut, (2) tool penetrating the workpiece, (3) complete chamber teeth and (4) beginning of the reversal of the spindle.

3. Results and Discussion

3.1. Film Formation on Ti6Al4V

The formation of the lubrication layer was calculated according to the equations of the regression line abstracted from the calibration curve of each polyol ester cutting fluid under study (Table 5). The *REO* values were calculated with the IRRAS spectra on the Ti6Al4V panels and the total organic matter adhered to the surface, allowing the quantification of the lubrication layer formation.

Table 5. The *REO*, *TOC*, and equations of the regression line results of the cutting fluid formulated with polyol esters.

MWF	*REO*	*TOC* (mg C)	Equations of the Regression Line % ester = *A* + *REO*·*B*			% ester (*w*/*w*)	Organic Matter (μmol)	Ester (μmol)
			A	B	R^2			
C18:1 × 1	0.074	0.557	1.523	284.966	1.000	22.564	1.17	0.56
C18:1 × 2	0.219	0.860	2.077	279.079	0.999	63.166	0.80	1.17
C18:1 × 3	0.231	1.344	2.403	259.594	0.995	62.481	1.28	1.23
C18:1 × 4	0.304	1.200	2.474	250.050	0.999	78.408	0.83	0.87

Figure 4 shows the results of the organic film formed on the Ti6Al4V surface and the corresponding amount of ester after the panel was dipped in the O/W emulsions. The addition of esters in the cutting fluid resulted in lubrication layer growth. The ability to increase the film formation can be attributed to the stronger adsorption of polar functional groups on the metal surface. The molecular structure of such esters has a clear influence on the layer formed. Noteworthy results were obtained with C18:1 × 3. When the strip was dipped in the trimethylolpropane trioleate emulsion, a higher amount of organic matter was adhered, including both ester and surfactants.

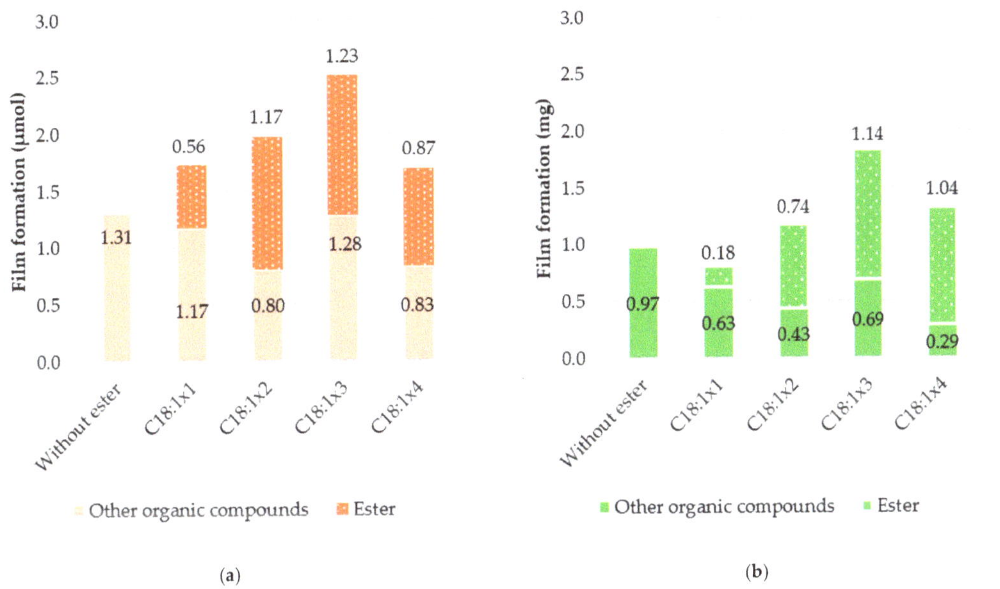

Figure 4. The film formation on the metal strips after the immersion of the panels in the cutting fluid: (**a**) in micromoles and (**b**) in milligrams.

At the same time, the presence of the ester interfered with the ability of other additives to attach to the metal surface. The results suggest that there is an adsorption competition on the surface between the surfactant and the ester molecules. The large molecular volume and polarity of pentaerythrityl tetraoleate prevents other molecules from being deposited on the metal, reducing the total organic matter. Therefore, it can be considered that the molecular structure of an ester in O/W emulsions has a substantial influence on its ability to form a lubricating film and on its conformation. The influence of the physico–chemical properties and the chemical composition, such as the branching structure and the polar functional groups on the film thickness has been observed by several authors [33], but unlike this study, they used neat esters.

When the Ti6Al4V strip was treated with C18:1 × 1 cutting fluid, the amount of ester that adhered to the surface was lower. This result can be attributed to the high stability of oil droplets in the emulsion, which are too stable to form a film. This relation between the oil droplet's stability and the film formation observed in this work was described by Ratoi-Salagean et al. [34], who correlated its incidence with a different emulsifier concentration.

3.2. *Influence of Polyol Esters in the Lubricity of Cutting Fluids*

To evaluate the lubricity of the cutting fluids, a tapping torque machine was used for experimental investigation. With this equipment, experiments are conducted by performing a tapping operation of a tool inside a hole with the cutting fluid and the torque is measured in situ. During the first tap, using deionized water as the cutting fluid, the tool welds in the Ti6Al4V workpiece. Comparing these results to the results from the cutting fluid without an ester, which showed that even after 15 taps the tool was not broken, it must be noticed that surfactants not only help to stabilize the emulsion, but they can also reduce the tapping torque. The lubricity properties of the surfactants observed in the present work were described by Benedicto et al. [35] who correlated their performance to the molecular structure of surfactants. Moreover, the addition of a polyol ester in the cutting fluid increases its lubricity and can reduce the tool wear by up to 37%.

The effect of the fatty acid ester in O/W emulsions on lubricity and wear is depicted in Figure 5 By comparing tapping torque values in the first tap, the ability of cutting fluids to lubricate can be studied. After the first tap, it is noted that the torque values increase with each tap due to tool wear.

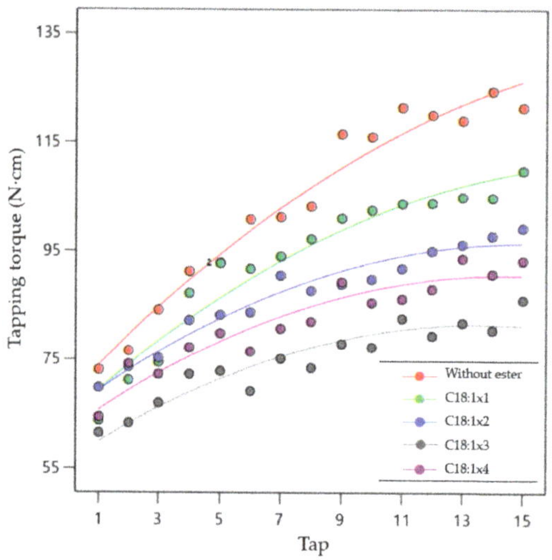

Figure 5. The effect of the addition of esters into an O/W emulsion on the tapping torque.

In order to analyze the effect of the cutting fluid on the tapping torque, a mathematical model was predicted by the design of experiments method (DOE). For each cutting fluid, a second order polynomial response has been fitted into the following equation:

$$Y = a + b \cdot X + c \cdot X^2, \quad (2)$$

where Y is the corresponding response (tapping torque) and X is the tap number.

The regression equations that have been obtained are collected in Table 6. The terms of these equations can be correlated with the lubricity performance (coefficient a) and the increase in tapping torque values in each consecutive tap due to the tool wear (coefficient b).

Table 6. The regression equations for a quadratic model.

Cutting Fluid	Regression Line
Without ester	$68.07 + 5.91\ X - 0.14\ X^2$
C18:1 × 1	$64.41 + 5.04\ X - 0.14\ X^2$
C18:1 × 2	$65.31 + 4.11\ X - 0.14\ X^2$
C18:1 × 3	$56.40 + 3.69\ X - 0.14\ X^2$
C18:1 × 4	$61.90 + 3.94\ X - 0.14\ X^2$

The analysis of variance (ANOVA) with a *F*-value of 202.82 implies that the model is significant and there is only a 0.01% chance that an *F*-value this large could occur due to noise. *p*-values less than 0.0500 indicate the model terms are significant. Additionally, the verification of the model was analyzed by R2, whose value is 0.969, very near to 1.

The results show a positive correlation between the lubricity and the total organic matter adhered and the amount of ester in the film layer. The ester was adsorbed to the metal surface with the fatty acid hydrocarbon end facing away from the metal surface, thus allowing a monolayer film formation. This is in accordance with the lubrication mechanism of the castor oil-in-water fluid described by Ni et al. [36]. Therefore, the fatty acid chain provided a sliding surface that reduced friction and facilitated chip evacuation through the tool's channels.

The wear protection was improved by increasing the ester, which resulted in stronger adsorption on the metal surface [37]. Remarkable results were obtained using C18:1 × 3. This can be attributed to the higher lubrication film that was formed on the metal surface, thus increasing lubricity and protecting the tool from wear. On the contrary, from the esters tested, C18:1 × 1 had the highest tool wear, corresponding to the polyol ester with the lowest ability to adhere on the Ti6Al4V surface. C18:1 × 3 improved the lubricity performance by 12.4% and decreased the tapping torque of consecutive taps by reducing the tool wear by 26.8% compared to the cutting fluid C18:1 × 1.

In comparison with C18:1 × 2, a slight decrease in the tapping values was detected using the cutting fluid C18:1 × 4. The molecule packing of C18:1 × 4 improves the deposition of the ester on the surface (1.04 mg of pentaerythrityl tetraoleate), increasing the lubricity properties of the cutting fluid. In terms of tool wear, the lowest indices were achieved with C18:1 × 4 fluid.

The results in this work demonstrate considerable potential for the introduction of the polyol ester as an alternative to mineral oil in water-based cutting fluids. Moreover, the associated environmental and health benefits of polyol esters make them more attractive substitutes. The characteristics of esters, such as their high biodegradability and the reduction in fossil carbon sources, reduces their contribution to global warming compared to mineral oil by approximately four times [38].

4. Conclusions

This study provides an evaluation of the influence of polyol esters in oil-in-water (O/W) emulsions on titanium alloys to overcome the challenges in formulating sustainable water-based cutting fluids. The emulsions were formulated by adding polyol esters with different molecular structures and surfactants to stabilize the O/W emulsion. The film formation on the surface and the tapping torque were measured to identify the factors that have a significant effect on lubricity performance.

Based on the results of the present experimental and analytical investigations, the following conclusions can be drawn:

- The molecular structure of esters influences the amount of the ester that adheres to the surface forming a lubricant film, which protects the tool from wear. By the addition of 1 mmol/L of ester in the water-based cutting fluid, trimethylolpropane trioleate ester can double the amount of ester adhered on the panel surface compared to isopropyl oleate.
- There is an adsorption competition on the surface between the ester and other additives. The molecular structure of esters has a high impact on the conformation of the lubricant film which, in turn, also has an impact on the lubricity properties of the cutting fluid.
- The increase in ester in the lubrication film improves the tribological performance and prolongs the tool life. The addition of a polyol ester in the cutting fluid increases the lubricity by up to 17% and can reduce the tool wear by up to 37%.
- From the polyol esters under study, C18:1 × 3 is preferred as a water-based cutting fluid for machining titanium alloys. It can double the amount of lubrication film

formed compared to the rest of the cutting fluids. The best lubricity results were obtained with C18:1 × 3. In terms of tool wear, the lowest rate was achieved with the trimethylolpropane trioleate O/W emulsion.

This study may be used to develop new sustainable and environmentally friendly cutting fluids for titanium alloys to replace conventional mineral oil water-based cutting fluids. However, further work needs to be conducted to investigate the industrial potential of these cutting fluids for titanium alloys, comparing them with other lubrication and cooling systems, such as cryogenic cooling.

Author Contributions: Conceptualization, E.B., E.M.R. and L.A.; methodology, E.B. and L.A.; validation, E.B. and L.A.; formal analysis, E.B., E.M.R., L.A. and M.A.S.-N.; investigation, E.B.; resources, E.B.; data curation, E.B.; writing—original draft preparation, E.B.; writing—review and editing, E.B., E.M.R. and L.A.; supervision, E.M.R.; project administration, E.M.R.; funding acquisition, E.M.R. All authors have read and agreed to the published version of the manuscript.

Funding: This research was supported by the Research Group of the UNED "Industrial Production and Manufacturing Engineering (IPME)", which has been financed in part by the to the Spanish Ministry of Science, Innovation and Universities (Project RTI2018-102215-B-I00), to the Industrial Engineering School-UNED (REF2021-ICF04) and to the Master in Advanced Manufacturing Engineering.

Data Availability Statement: Not applicable.

Acknowledgments: The authors thank the technical support provided by Lluis Beltrán and Angel Navarro, from Industrial Química Lasem, and by the Research Group of the UNED "Industrial Production and Manufacturing Engineering (IPME)".

Conflicts of Interest: The authors declare no conflict of interest. The funders had no role in the design of the study; in the collection, analysis, or interpretation of data; in the writing of the manuscript; or in the decision to publish the results.

References

1. Shokrani, A.; Dhokia, V.; Newman, S. Environmentally conscious machining of difficult-to-machine materials with regard to cutting fluids. *Int. J. Mach. Tools Manuf.* **2012**, *57*, 83–101. [CrossRef]
2. Sagin, S.V.; Solodovnikov, V.G. Estimation of Operational Properties of Lubricant Coolant Liquids by Optical Methods. *Int. J. Appl. Eng. Res.* **2017**, *12*, 8380–8391.
3. Madanchi, N.; Kurle, D.; Winter, M.; Thiede, S.; Herrmann, C. Energy Efficient Process Chain: The Impact of Cutting Fluid Strategies. *Procedia CIRP* **2015**, *29*, 360–365. [CrossRef]
4. Castellanos, S.D.; Cavaleiro, A.J.; Jesus, A.M.P.D.; Neto, R.; Alves, J.L. Machinability of titanium aluminides: A review. *Proc. Inst. Mech. Eng. Part L J. Mater. Des. Appl.* **2019**, *233*, 426–451. [CrossRef]
5. Machado, R.; Wallbank, J. Machining of Titanium and its Alloys—A Review. *Proc. Inst. Mech. Eng.* **1990**, *204*, 53–60. [CrossRef]
6. Younas, M.; Husain Imran Jaffery, S.; Khan, M.; Ahmad, R.; Ali, L.; Khan, Z.; Khan, A. Tool Wear Progression and its Effect on Energy Consumption in Turning of Titanium Alloy (Ti-6Al-4V). *Mech. Sci.* **2019**, *10*, 373–382. [CrossRef]
7. Shaikh, V.A.; Boubekri, N. Using vegetable-oil-based sustainable metal working fluids to promote green manufacturing. *Int. J. Manuf. Mater. Mech. Eng.* **2020**, *10*, 1–19. [CrossRef]
8. Chetan; Ghosh, S.; Venkateswara Rao, P. Application of sustainable techniques in metal cutting for enhanced machinability: A review. *J. Clean. Prod.* **2015**, *100*, 17–34. [CrossRef]
9. Pervaiz, S.; Kannan, S.; Kishawy, H.A. An extensive review of the water consumption and cutting fluid based sustainability concerns in the metal cutting sector. *J. Clean. Prod.* **2018**, *197*, 134–153. [CrossRef]
10. Pimenov, D.Y.; Mia, M.; Gupta, M.K.; Machado, A.R.; Tomaz, Í.V.; Sarikaya, M.; Wojciechowski, S.; Mikolajczyk, T.; Kaplonek, W. Improvement of machinability of Ti and its alloys using cooling-lubrication techniques: A review and future prospect. *J. Mater. Res. Technol.* **2021**, *11*, 719–753. [CrossRef]
11. Pereira, O.; Martín-Alfonso, J.E.; Rodríguez, A.; Calleja, A.; Fernández-Valdivielso, A.; López de Lacalle, L.N. Sustainability analysis of lubricant oils for minimum quantity lubrication based on their tribo-rheological performance. *J. Clean. Prod.* **2017**, *164*, 1419–1429. [CrossRef]
12. Ozimina, D.; Kowalczyk, J.; Madej, M.; Nowakowski, Ł.; Kulczycki, A. The impact of the type of cutting fluid on the turning process. *Tribologia* **2017**, *3*, 119–126. [CrossRef]
13. Prasad, K.; Chakraborty, S. Application of the modified similarity-based method for cutting fluid selection. *Decis. Sci. Lett.* **2018**, *7*, 273–286. [CrossRef]
14. Yang, Y.; Zhang, C.; Wang, Y.; Dai, Y.; Luo, J. Friction and wear performance of titanium alloy against tungsten carbide lubricated with phosphate ester. *Tribol. Int.* **2016**, *95*, 27–34. [CrossRef]

15. Yan, P.; Rong, Y.; Wang, G. The effect of cutting fluids applied in metal cutting process. *Proc. Inst. Mech. Eng. Part B J. Eng. Manuf.* **2016**, *230*, 19–37. [CrossRef]
16. Pedišić, L.; Polenus, I. Advantages of Biodegradable Fluids Application At Metalworking Operations. In Proceedings of the 16th International Research/Expert Conference Trends in the Development of Machinery and Associated Technology, TMT 2012, Dubai, United Arab Emirates, 10–12 September 2012; pp. 43–46.
17. Assenova, E.; Majstorovic, V.; Vencl, A.; Kandeva, M. Green Tribology and Quality of Life. *Int. J. Adv. Qual.* **2012**, *40*, 1–6.
18. Winter, M.; Thiede, S.; Herrmann, C. Influence of the cutting fluid on process energy demand and surface roughness in grinding—A technological, environmental and economic examination. *Int. J. Adv. Manuf. Technol.* **2015**, *77*, 2005–2017. [CrossRef]
19. Yang, Y.; Zhang, C.; Dai, Y.; Luo, J. Tribological properties of titanium alloys under lubrication of SEE oil and aqueous solutions. *Tribol. Int.* **2017**, *109*, 40–47. [CrossRef]
20. Zulkifli, N.W.M.; Azman, S.S.N.; Kalam, M.A.; Masjuki, H.H.; Yunus, R.; Gulzar, M. Lubricity of bio-based lubricant derived from different chemically modified fatty acid methyl ester. *Tribol. Int.* **2016**, *93*, 555–562. [CrossRef]
21. Pettersson, A. High-performance base fluids for environmentally adapted lubricants. *Tribol. Int.* **2007**, *40*, 638–645. [CrossRef]
22. Cecilia, J.A.; Ballesteros Plata, D.; Alves Saboya, R.M.; Tavares de Luna, F.M.; Cavalcante, C.L.; Rodríguez-Castellón, E. An Overview of the Biolubricant Production Process: Challenges and Future Perspectives. *Processes* **2020**, *8*, 257. [CrossRef]
23. Talib, N.; Rahim, E.A. Experimental evaluation of physicochemical properties and tapping torque of hexagonal boron nitride in modified jatropha oils-based as sustainable metalworking fluids. *J. Clean. Prod.* **2018**, *171*, 743–755. [CrossRef]
24. Canter, N. Formulating water-based MWFs in the 21st Century. *Tribol. Lubr. Technol.* **2017**, *73*, 42–55.
25. Debnath, S.; Mohan, M.; Sok, Q. Environmental friendly cutting fluids and cooling techniques in machining: A review. *J. Clean. Prod.* **2014**, *83*, 33–47. [CrossRef]
26. Wickramasinghe, K.C.; Sasahara, H.; Rahim, E.A.; Perera, G.I.P. Green Metalworking Fluids for sustainable machining applications: A review. *J. Clean. Prod.* **2020**, *257*, 120552. [CrossRef]
27. Ma, L.R.; Zhang, C.H.; Liu, S.H. Progress in experimental study of aqueous lubrication. *Chin. Sci. Bull.* **2012**, *57*, 2062–2069. [CrossRef]
28. Lancaster, J.K. A review of the influence of environmental humidity and water on friction, lubrication and wear. *Tribol. Int.* **1990**, *23*, 371–389. [CrossRef]
29. Madanhire, I.; Mbohwa, C. Environmentally Adapted Lubricants. In *Mitigating Environmental Impact of Petroleum Lubricants*; Springer: Berlin/Heidelberg, Germany, 2016; pp. 165–178.
30. Benedicto, E.; Carou, D.; Rubio, E.M.; Batlle, L. A novel method for the determination of fatty acid esters in aqueous emulsion on Ti6Al4V surface with IRRAS and carbon quantification. *Tribol. Int.* **2018**, *128*, 155–160. [CrossRef]
31. Pereira, I.C.; Da Silva, M.B.; Da Cunha, D.F.; Sales, W.F. Analysis of tapping process in three types of cast iron. *Int. J. Adv. Manuf. Technol.* **2016**, *82*, 1041–1048. [CrossRef]
32. Uzun, G.; Korkut, I. The effects of cutting conditions on the cutting torque and tool life in the tapping process for AISI 304 stainless steel. *Mater. Tehnol.* **2016**, *50*, 275–280. [CrossRef]
33. Mannekote, J.K.; Kailas, S.V.; Venkatesh, K.; Kathyayini, N. Environmentally friendly functional fluids from renewable and sustainable sources—A review. *Renew. Sustain. Energy Rev.* **2018**, *81*, 1787–1801. [CrossRef]
34. Ratoi-Salagean, M.; Spikes, H.A.; Rieffe, H.L. Optimizing Film Formation by Oil-in-Water Emulsions. *Tribol. Trans.* **1997**, *40*, 569–578. [CrossRef]
35. Benedicto, E.; Rubio, E.M.; Carou, D.; Santacruz, C. The Role of Surfactant Structure on the Development of a Sustainable and Effective Cutting Fluid for Machining Titanium Alloys. *Metals* **2020**, *10*, 1388. [CrossRef]
36. Ni, J.; Feng, K.; He, L.; Liu, X.; Meng, Z. Assessment of water-based cutting fluids with green additives in broaching. *Friction* **2020**, *8*, 1051–1062. [CrossRef]
37. Chan, C.H.; Tang, S.W.; Mohd, N.K.; Lim, W.H.; Yeong, S.K.; Idris, Z. Tribological behavior of biolubricant base stocks and additives. *Renew. Sustain. Energy Rev.* **2018**, *93*, 145–157. [CrossRef]
38. Ekman, A.; Börjesson, P. Life cycle assessment of mineral oil-based and vegetable oil-based hydraulic fluids including comparison of biocatalytic and conventional production methods. *Int. J. Life Cycle Assess.* **2011**, *16*, 297–305. [CrossRef]

metals

Article

Split-Plot I-Optimal Design Optimisation of Combined Oil-Based and Friction Stir Rotation-Assisted Heating in SPIF of Ti-6Al-4V Titanium Alloy Sheet under Variable Oil Pressure

Tomasz Trzepieciński [1,*], Marcin Szpunar [2] and Robert Ostrowski [3]

[1] Department of Manufacturing and Production Engineering, Faculty of Mechanical Engineering and Aeronautics, Rzeszow University of Technology, al. Powst. Warszawy 8, 35-959 Rzeszów, Poland
[2] Doctoral School of Engineering and Technical Sciences, Rzeszow University of Technology, al. Powst. Warszawy 12, 35-959 Rzeszów, Poland; d547@stud.prz.edu.pl
[3] Department of Materials Forming and Processing, Faculty of Mechanical Engineering and Aeronautics, Rzeszow University of Technology, al. Powst. Warszawy 8, 35-959 Rzeszów, Poland; rostrows@prz.edu.pl
* Correspondence: tomtrz@prz.edu.pl

Abstract: The aim of this paper is to determine the optimal input parameters for the process in order to ensure the maximum formable wall angle is obtained in a conical frustum with a varying wall angle fabricated using Single Point Incremental Forming (SPIF). The test material was 0.8-mm-thick Ti-6Al-4V titanium alloy sheets, and the test used a tungsten carbide tool with a rounded tip with a radius of 4 mm. Complete workpieces were heated using hot oil with a temperature of about 200 °C, and in addition, the high rotation speed of the forming tool generated an amount of friction heat. The input parameters were tool rotational speed, feed rate, step size, and tool rotation direction. Various oil pressures were used to improve both the accuracy of the components formed and the friction heating process. On the basis of calculations performed by means of the response surface methodology, split-plot I-optimal design responses were obtained by means of polynomial regression models. Models were fitted using REstricted Maximum Likelihood (REML), and *p*-values are derived using the Kenward–Roger approximation. Observation of the fracture surface of Ti-6Al-4V drawpieces showed that the destruction is as a result of ductile fracture mode. Tool rotational speed and step size are the most significant factors that affect the axial force, followed by feed rate. It was also found that step size is the most significant factor that affects the in-plane SPIF force.

Keywords: feed rate; incremental sheet forming; SPIF; single point incremental forming; tool rotational speed

Citation: Trzepieciński, T.; Szpunar, M.; Ostrowski, R. Split-Plot I-Optimal Design Optimisation of Combined Oil-Based and Friction Stir Rotation-Assisted Heating in SPIF of Ti-6Al-4V Titanium Alloy Sheet under Variable Oil Pressure. *Metals* **2022**, *12*, 113. https://doi.org/10.3390/met12010113

Academic Editors: Guanyu Deng, Hongtao Zhu and Anh Kiet Tieu

Received: 15 December 2021
Accepted: 5 January 2022
Published: 7 January 2022

1. Introduction

Single-point incremental forming (SPIF), proposed by Iseki [1], is a flexible method of sheet metal forming that does not require the use of dies adapted to the shape of the product. The SPIF method and its variation, two-point incremental forming (TPIF), can be used to produce products with complex shapes, both convex and concave, using shank tools with rounded tips. These tools move along the programmed trajectory, gradually sinking into the formed sheet [2,3]. A milling machine or a numerically controlled lathe with at least three controlled axes is required to operate the SPIF process. Incremental sheet forming (ISF) methods that use a robotic arm to drive the working spindle have also been developed.

Due to the point character of the contact of the tool with the deformed sheet, the processing speed in this method is lower than in conventional sheet metal forming (SMF) methods that require the production of a die and punch [4]. However, die-less ISF is more effective when machining prototype parts, small-series components, and highly customised parts such as medical implants. Due to the local nature of plastic deformation, the deformation achievable in SPIF is greater than with conventional SMF. Comparison of the conventional SMF process intended to form a pyramid-shaped part with SPIF can be found in a paper by Petek et al. [5]. Over the years, many varieties of SPIF have been proposed for forming hard-to-deform alloys in cold forming conditions such as warming sheet material using hot air blowers [6] or using a furnace [7], electrically assisted ISF [8,9], laser-assisted ISF [10,11], and friction-assisted ISF [12].

During the forming process, due to the elastic–plastic deformation conditions, the forming process is prone to instability and the part is subject to springback [13,14] and pillow effect [15–17]. After forming, skirt springback occurs due to the elastic recovery of the sheet. Elimination of shape errors is achieved by optimisation of the tool trajectory in order to compensate for the elastic recovery of the sheet material. There are many studies designed to increase the accuracy of ISF carried out in cold and hot forming conditions. The process conditions of SPIF depend on many parameters that can interactively affect the surface finish, shape and dimensional accuracy of the drawpieces, wall thickness and the possibility of producing large sheet deformations. These parameters include: tool rotational speed, tool diameter, feed rate, step depth and lubrication conditions. Numerous investigations designed to optimise forming conditions using Analysis of Variance (ANOVA), Taguchi design of experiment, regression analysis, Non-dominated Sorting Genetic Algorithm (NSGA) and Response Surface Methodology (RSM) can be found in the literature. Honarpisheh et al. [18] studied the interaction and main effect of the process parameters viz. rotational speed, tool diameter, step down, on the wall and fracture depth in SPIF of 1050 aluminium alloy sheets using the ANOVA method and a multi-response optimisation. Kurra et al. [19] investigated SPIF of extra deep drawing steel sheets using the Box–Behnken design. They evaluated the effect of process parameters on surface roughness and manufacturing time using ANOVA, NSGA and RSM approaches. Szpunar et al. [20] developed the mathematical relations between feed rate, pitch, spindle speed and surface roughness on the inner surfaces of Grade 2 titanium drawpieces defined using the RSM and Central Composite Design (CCD). It was found that step size directly affects the surface roughness on the inner surface of the component. Truncated cone specimens formed with high values of tool rotational speed showed poor surface qualities. Vahdani et al. [9] employed Design of Experiment (DoE) and ANOVA to study the formability of Ti-6Al-4V titanium alloy sheet. The results showed that SPIF of the study material can be improved by electric hot incremental sheet forming. The electric current and the lubricant type have a significant effect on the maximum achievable forming depth. Najm and Paniti [21] used artificial neural networks to explore and estimate the relative importance of SPIF parameters during forming of AlMn1Mg1 aluminium alloy sheets. One of the key findings is that tool characteristics (surface roughness, tool tip shape and tool material) play an essential role in the accuracy of the final component. Maji and Kumar [22] developed RSM and an Adaptive Neuro-Fuzzy Inference System (ANFIS) with a non-dominated sorting genetic algorithm to predict the outcome of SPIF components. They found that surface roughness (Ra) was most significantly affected by step height. Moreover, the formable wall angle and deformed sheet thickness were mostly affected by feed rate. Grün et al. [12] applied one-factor-at-a-time (OFAT) and DoE methods to investigate the strength of the correlations between the kinematics of the tool and the formability of the Ti6Al4V sheets in friction stir ISF. Tool rotation speed was found to be the dominant friction stir SPIF parameter. Step depth and feed rate had a minor impact.

There are other reports investigating process parameters directly without using any statistical procedures. Durante et al. [23] studied the influence of tool rotation in SPIF of 7075-T0 aluminium alloy pyramid frustums in terms of forming forces, surface roughness and forming temperature. Temperature measurement highlights different heating of the sheet and in particular a proportional dependence on the speed of rotation because of the relative motion between the sheet and tool. Hussain et al. [24] studied the effect of feed rate, tool diameter, pitch, and friction conditions on the formability of a commercially pure (CP) titanium sheet varying wall angle conical frustum (VWACF). It was found that friction at the tool/blank interface does not affect the formability of CP titanium sheet. A decrease in the feed rate increases the formability. Racz et al. [25] chose the most highly recommended incremental forming process for Ti-6Al-4V titanium alloy cranioplasty plates using a decision-making method based upon AHP (analytic hierarchy process). They concluded that the accuracy of the wall angle was not significantly influenced by the tool diameter and toolpath strategy. This can be explained by the fact that the low plasticity of the Ti-6Al-4V titanium alloy does not lead to a significant degree of springback. Mohanraj and Elangovan [26] investigated the effect of process parameters such as tool rotational speed, feed rate, step depth and tool diameter on the geometrical accuracy and thinning in SPIFed Ti-6Al-4V aerospace components both experimentally and using the finite element method (FEM). Palumbo et al. [27] formed Ti-6Al-4V truncated cones with the aim of investigating the effect of both the tool diameter/pitch ratio (D/p) and the wall angle on the formability of the material. An analysis of shape errors, and thinning measurements highlighted the fact that the parameter D/p plays a key role in SPIF. Naranjo et al. [28] conducted a numerical campaign to study the formability of Ti-6Al-4V VWACF in a warm SPIF process. The effect of temperature on formability was experimentally evaluated in terms of the maximum achieved wall angle. Increase in temperature improved formability however the temperature did not significantly influence the surface finish of drawpieces. The effects of the forming temperature on material failure and more particularly on geometric accuracy while forming a Ti-6Al-4V denture prosthesis were investigated by Sbayti et al. [29]. The numerical models conducted showed that SPIF of a denture base made of the one of the most frequently used alloys for biomedical applications has the potential for real medical applications. In another paper, Sbayti et al. [30] numerically investigated the effect of tool diameter and temperature on the failure and geometric accuracy of the acetabular component of a hip prosthesis made of Ti-6Al-4V titanium alloy. The emerging trends and development of SPIF in titanium and its alloys can be found in review papers by Oleksik et al. [31] and Trzepieciński et al. [32].

In this manuscript, an I-optimal split-plot algorithm was used to build a response surface design for SPIF of Ti-6Al-4V alloy. Due to the poor room-temperature formability of a Ti-6Al-4V alpha-beta titanium alloy, deformation under high temperature is needed [29]. The main reason for its poor formability is low ductility at room temperature due to the hexagonal close-packed (HCP) structure in α-phase. To overcome this limitation, the SPIF experiments were conducted in warm conditions. Several authors have developed systems for heating the sheet at the tool contact point by the Joule effect [8,9], friction stir heating [12], laser-assisted heating [10,11], or by employing heat bands [33]. The non-uniform heating drawback of these methods and poor efficiency of the heat bands method is overcome in this paper by using combined oil-based and friction stir rotation-assisted heating of the workpiece. The input parameters were tool rotational speed, feed rate, step size, and tool rotation direction. To further improve sheet formability, a variable oil pressure in the die cavity is considered. The mathematical relations are defined from the response surfaces to predict both the maximum formable wall angle of the conical frustum with a varying wall angle and the components of the forming force.

2. Materials and Methods

2.1. Material

Ti-6Al-4V titanium alloy sheet metal 0.8 mm thick was used as the test material. At room temperature the microstructure of Ti-6Al-4V alloy mainly consists of hexagonal closed packed (HCP) for the α and body centred cubic (BCC) for the β phases, respectively [34,35]. This two-phase (α-β) Ti-based alloy is the most frequently used Ti-based alloy in a variety of applications, including blades, discs and rings of turbines and constructional elements of airplanes. Ti-6Al-4V material is also commonly used as biocompatible implants in medicine applications. It has high corrosion resistance, good weldability, excellent strength, and low modulus of elasticity. However, due to its complex microstructure at room temperature, this alloy is commonly formed at elevated temperatures. The chemical composition of Ti-6Al-4V alloy is given in Table 1 according to ISO 5832/3 [36].

Table 1. Chemical composition of Ti-6Al-4V (wt.%).

Al	V	Fe	O	C	N	H	Ti
5.5	3.5	<0.3	<0.2	<0.08	<0.05	<0.0015	remainder

2.2. Experimental Setup

The tests for forming the drawpieces under combined oil-based and friction stir rotation-assisted heating in warm SPIF were carried out with the use of a die consisting of a housing and a blankholder ensuring the fixing of the displacement of the workpiece flange. Structural diagram and view of the experimental setup are shown in Figure 1a,b, respectively. The sheet metal clamp was secured by means of a set of screws. An electric oil heater (Tempco Electric Heater Corp., Wood Dale, IL, USA) was placed in the die cavity and the oil pressure was kept constant during forming by means of a valve. The forming device was mounted on the bed of a PS95 vertical CNC milling machine (Makino Milling Machine Co. Ltd., Tokyo, Japan). A workpiece with a diameter (TIMET Titanium Metals Corp., Toronto, OH, Canada) of 100 mm was used to form conical frustums with varying wall angles (Figure 2). A tungsten carbide tool (HHT – Hartmetall GmbH & Co. KG, Dornstadt, Germany) with a rounded tip with a radius of 4 mm was mounted in the face mill through an ER32 collet chuck (Haimer GmbH, Igenhausen, Germany). A grease-free dry anti-friction spray of MoS_2 (WEICON GmbH & Co. KG, Münster, Germany) was used in the experiments. This lubricant is pressure and temperature resistant from −185 °C to 400 °C. Surfaces were cleaned and degreased prior to application.

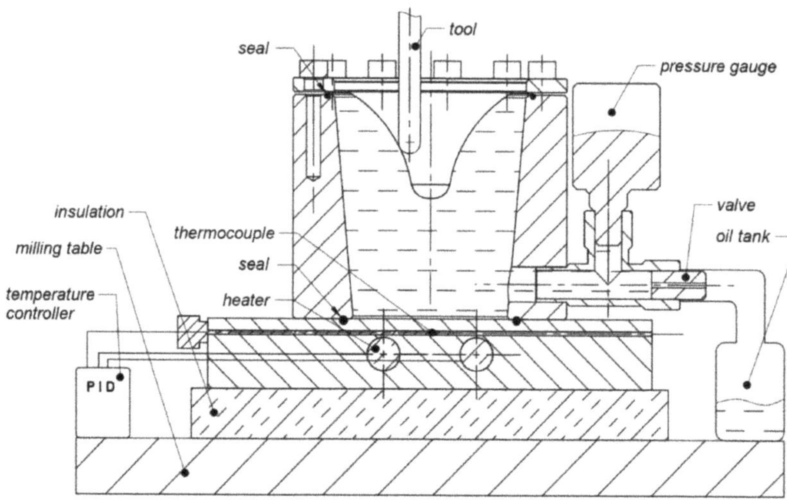

(a)

Figure 1. *Cont.*

(b)

Figure 1. (a) Structural diagram and (b) view of experimental setup.

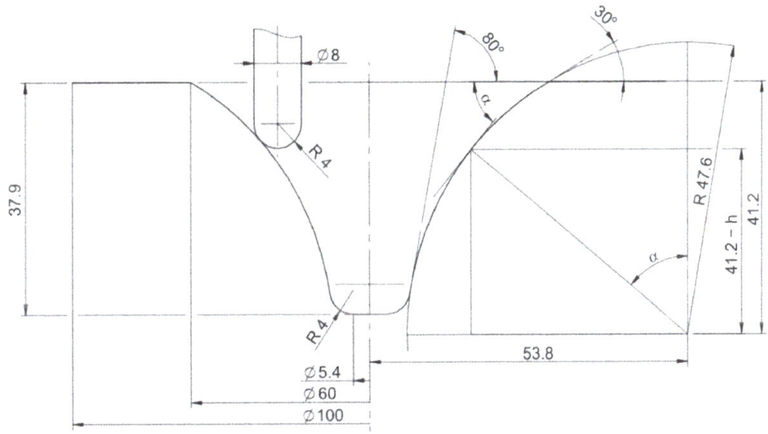

Figure 2. Geometry and dimensions of varying wall angle conical frustums (dimensions in mm).

The experiments stopped after fracture of the part. Height of drawpieces h was measured using height gauge. The maximum formable wall angle α was determined based on the drawpiece height h:

$$\alpha = \arccos\frac{41.2 - h}{47.6} \tag{1}$$

The axial force F_z and the horizontal components of the forming force F_x and F_y were measured by a high-accuracy piezoelectric dynamometer (Kistler Holding AG, Winterthur, Switzerland) with a maximum sample rate per channel of 200 kHz. Based on the two horizontal components of the forming force F_x and F_y, the in-plane force F_{xy} was determined according to the formula:

$$F_{xy} = \sqrt{F_x^2 + F_y^2} \tag{2}$$

The tool indented into the workpiece followed a spiral path. The tool trajectory (Figure 3) was generated using NX CAM software (version 1938, Siemens Digital Industries Software, Plano, TX, USA) based on the numerical model of the desired shape of the drawpiece.

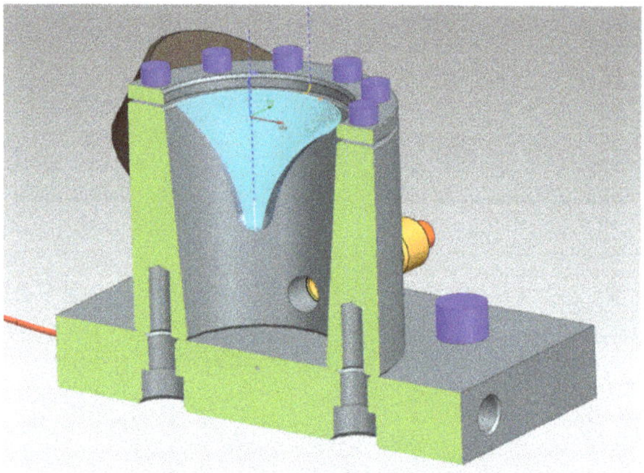

Figure 3. Toolpath trajectory generated in NX CAM software.

2.3. Plan of Experiments

Split-plot I-optimal design (Design Expert, version 12, Stat-Ease Inc., Minneapolis, MN, USA) was used to determine the input parameters of the SPI process in order to maximise the wall angle and to minimise the in-plane and axial forming force components. Design of industrial experiments with complete randomisation is limited due to cost constrains. The split-plot design, which involves a restricted randomisation, often provides a reasonable alternative [37]. Split-plot designs were originally used in agriculture where plots of land were subdivided in relatively large portions known as whole plots. Each of the levels of the whole-plot factors were then randomly assigned to these plots. Whole plots were further divided into smaller portions known as subplots, to which subplot factors were applied [37]. In split-plot design, the hard-to-change factors act as whole-plot factors and easy-to-change factors belong to sub-plots. I-optimal designs are noted for their integrated variance. In I-optimal design, the computational model adjusts the response surface that determines the optimal conditions, by minimising the mean differences (variances) of the predicted numerical data of the independent variables. The model assumes the forecast of missing combinations of independent variables of the experiment using the actual plan of the experiment used for optimisation [38].

Initial range of input parameters for split-plot I-optimal design was determined in the preliminary experimental studies of VWACF forming. A reasonable range of parameters was determined considering the following problems. A rotational speed of the tool that was too high caused excessive heating not only of the sheet but also of the tool. Therefore, excessive build-up was observed in the tool tip surface. The maximum value of the feed rate (2000 mm/min) was limited by the inertia of the MHWT (machine–holder–workpiece–tool) system, causing strong vibrations when tool moving along a trajectory with small radius. In addition, the feed must be considered together with the rotational speed to ensure proper friction-assisted heating of the sheet metal. The device is structurally adjusted (sealing, valve type) to work under a maximum oil pressure of 4 bar. Step size greater than 0.5 mm caused premature cracking of the drawpieces due to insufficient local heating of the material by tool interaction.

The predominant input factors and range of their variation, which have most influence on the maximum formable wall angle and components of the forming force, were identified from preliminary experiments. The input parameters were oil pressure p, tool rotational speed n, feed rate f, step size a_p and direction of tool rotation (Table 2). The last parameter was selected based on the study of Szpunar et al. [20], who found that the direction of tool rotation in relation to the feed direction (anticlockwise and clockwise—Figure 4) influenced the possibility of receiving drawpieces without the risk of cracking.

Table 2. Factors and levels used in the split-plot I-optimal design *.

SPIF Parameter	Factor	Unit	Change	Low Level	High Level
Oil pressure p	a	bar	hard	1	4
Tool rotational speed n	B	rpm	easy	100	1000
Feed rate f	C	mm/min	easy	500	2000
Step size a_p	D	mm	easy	0.1	0.5
Tool rotation direction	E	-	easy	−	+

* "+"—clockwise direction, "−"—counterclockwise direction.

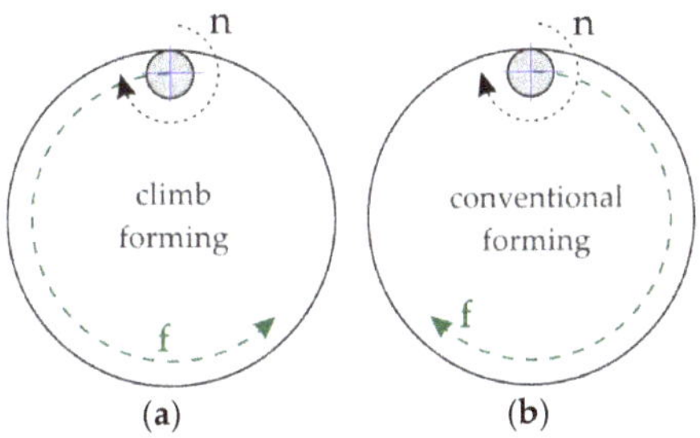

Figure 4. Spindle rotation: (**a**) anticlockwise; and (**b**) clockwise.

The split-plot I-optimal design was composed of five levels, and 25 experiments (Table 3) were carried out to optimise the input variables. The points of the split-plot I-optimal design in the workspace are presented in Figure 5.

Table 3. Plan of experiments for split-plot I-optimal design ("+"—clockwise direction, "−"—counterclockwise direction of tool rotation).

Number of Experiment	Oil Pressure, Bar	Tool Rotational Speed, rpm	Feed Rate, mm/min	Step Size a_p, mm	Direction of Tool Rotation
1	2	1000	1475	0.1	−
2	2	100	2000	0.1	−
3	2	514	1370	0.31	+
4	2	1000	500	0.1	+
5	2	590	1175	0.35	−
6	2	100	2000	0.5	+
7	2	100	500	0.5	−
8	2	122	1108	0.1	+
9	2	1000	1145	0.33	+
10	2	559	2000	0.26	−
11	4	100	2000	0.3	+
12	4	595	620	0.5	+
13	4	100	883	0.14	−
14	4	757	1550	0.1	+
15	4	1000	2000	0.37	−
16	1	581	538	0.5	+
17	1	550	500	0.1	−
18	1	1000	2000	0.5	−
19	1	100	1423	0.32	−
20	1	762	2000	0.12	+
21	3	1000	2000	0.5	+
22	3	100	500	0.27	+
23	3	680	500	0.2	−
24	3	1000	500	0.5	−
25	3	343	1565	0.5	−

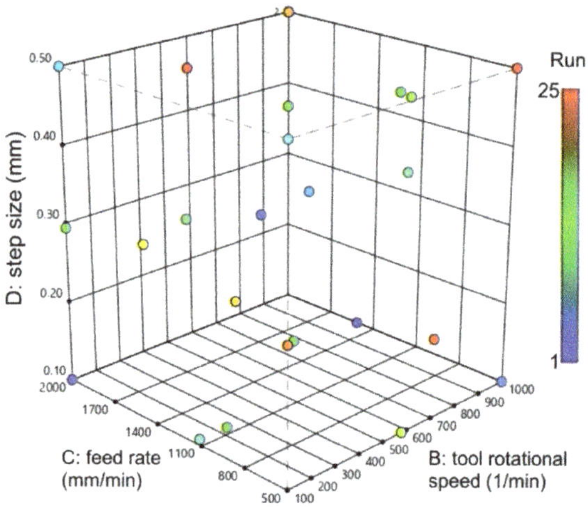

Figure 5. Points of the split-plot I-optimal design in the workspace.

3. Results and Discussion

3.1. Split-Plot I-Optimal Design with RSM

Analysis of variance was used to determine the relationships between the input variables and the specific output variable: maximum formable wall angle, in-plane forming force F_{xy} and axial forming force F_z (Figure 6). Analysis of variance explains the probability with which the selected factors may be the cause of differences between the observed group means. The aim of the statistical analysis was to determine the significance of the influence of individual parameters on the selected output variables using Fisher's test at a significance level $\alpha = 0.05$. Based on the statistical tables for F (0.05), an analysis of the significance of individual variables was performed. On the basis of the calculations performed by means of the RSM, optimal responses were obtained by means of polynomial regression models. Models are fitted using REstricted Maximum Likelihood (REML) and *p*-values are derived using the Kenward–Roger approximation. Table 4 shows the results of the experimental tests based on the split-plot I-optimal design.

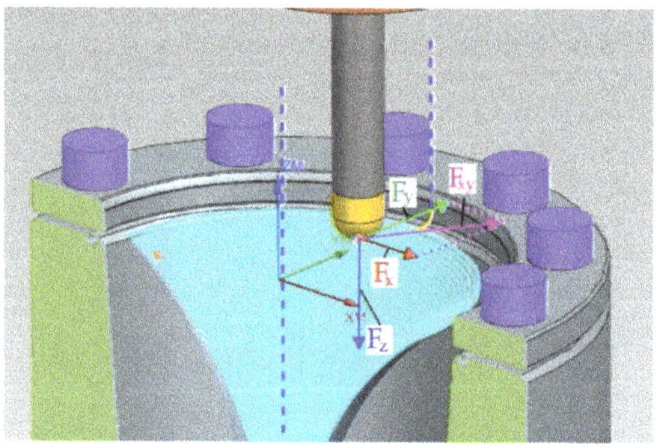

Figure 6. Force components in SPIF.

Table 4. Results of the SPIF forming.

Number of Experiment	Maximum Formable Wall Angle α, °	Maximum Axial Force F_z, N	Maximum in-Plane Force F_{xy}, N
1	60.9	1536	476
2	48.2	1621	644
3	55	2335	849
4	65.1	1132	148
5	62.9	2070	856
6	53.8	2560	874
7	54.6	2566	966
8	51.7	1633	411
9	56.1	1805	752
10	48.1	2097	669
11	52.2	2360	757
12	52.7	2043	794
13	53.8	1831	641
14	52.3	1573	580
15	66	2020	711
16	42.3	2377	435
17	44.8	1275	132
18	45	2289	633
19	44	2159	572
20	40.6	1409	147
21	43.3	2346	501
22	40.3	2115	308
23	52	1663	310
24	65.5	1819	556
25	45.7	2464	735

The experiments stopped after fracture of the part. The highest wall angle $\alpha = 66°$ was obtained for run no. 15. The drawpiece fracture was observed on the side surface of the cone slightly below the drawpiece bottom (Figure 7). The greater the wall angle, the greater the tensile stresses in the drawpiece wall in the direction of downward movement of the tool. Simultaneously with the increase in value of the wall angle, the share of circumferential stresses acting on the edge of the bottom of the drawpiece decreases. Figures 8 and 9 show SEM micrographs of a fracture surface of Ti-6Al-4V VWACF. Observation of the fracture surfaces of Ti-6Al-4V drawpieces showed that the destruction is a result of the ductile fracture mode (Figures 8c and 9b). Ductile cracking occurs by nucleation and void growth and usually begins with particles of a different phase [39]. During ductile fracture, the formation and joining of cracks takes place due to the plastic flow of the workpiece material. An underside view of the fracture surface, related to the inner surface of the sheet, shows dimples similar to the ones due to ductile fracture under tensile stress [40]. Microcracks in the subsurface layer are observed on the inner surface of the drawpiece (Figure 8b). Due to the severe impact of the tool tip, the formability of the material in the subsurface layer in contact with the tool has decreased. As a result of the reduction in formability, the material in the subsurface layer was susceptible to fracture. The outer part of the fracture surface (the part in contact with the outer surface of the drawpiece), presents small, partially formed dimples typical of shear load conditions (Figures 8a and 9d) [40]. This can be explained by various stress states. On the inner side of the part, the tip of the tool pushes the material downward making tensile stress predominant, while, on the outer side, the circumferential movement of the bottom tool and the related shear effect generate shear-type surface fracture [41]. As mentioned, the fracture is first stretched due to the meridional tensile stresses due to the downward movement of the tool, and then expanded in the horizontal direction.

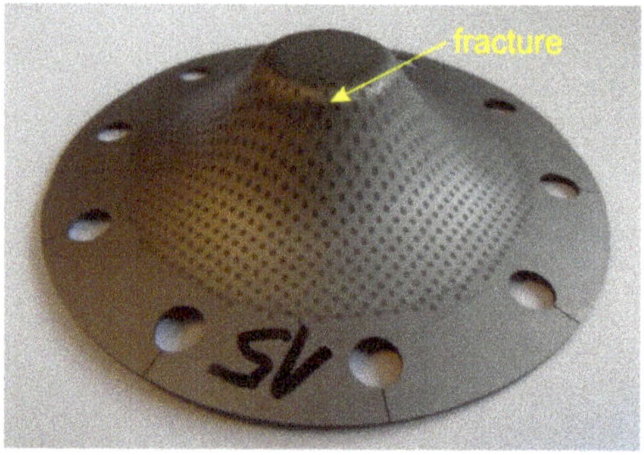

Figure 7. Fractured VWACF produced in run 15.

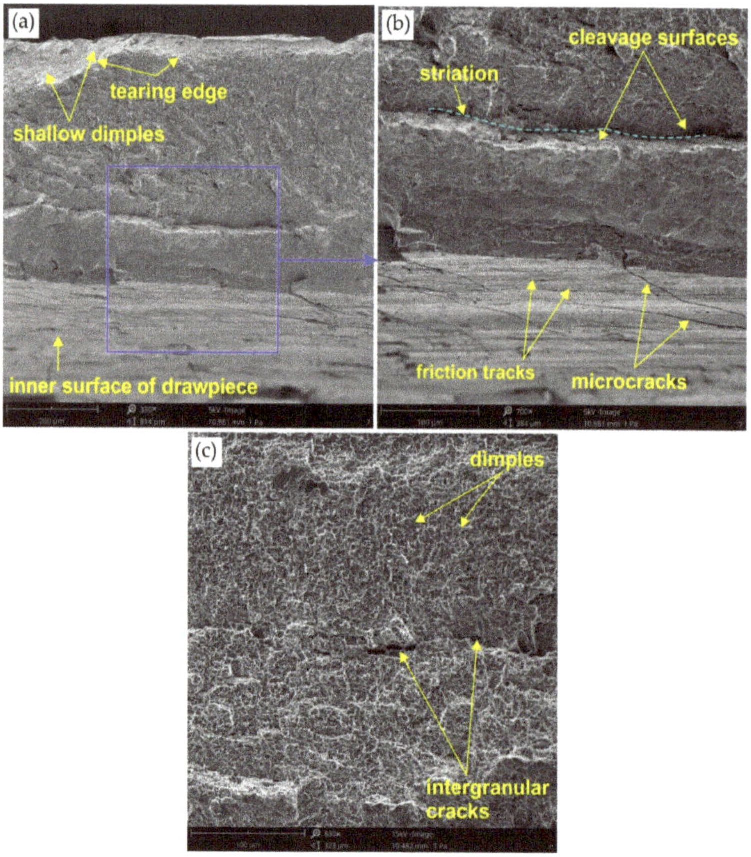

Figure 8. SEM micrographs of the fracture surface of VWACF produced in run 15: (**a**) cross-section of the fractured surface, (**b**) view of the near-edge inner surface of the drawpiece, (**c**) magnification of the ductile fracture area.

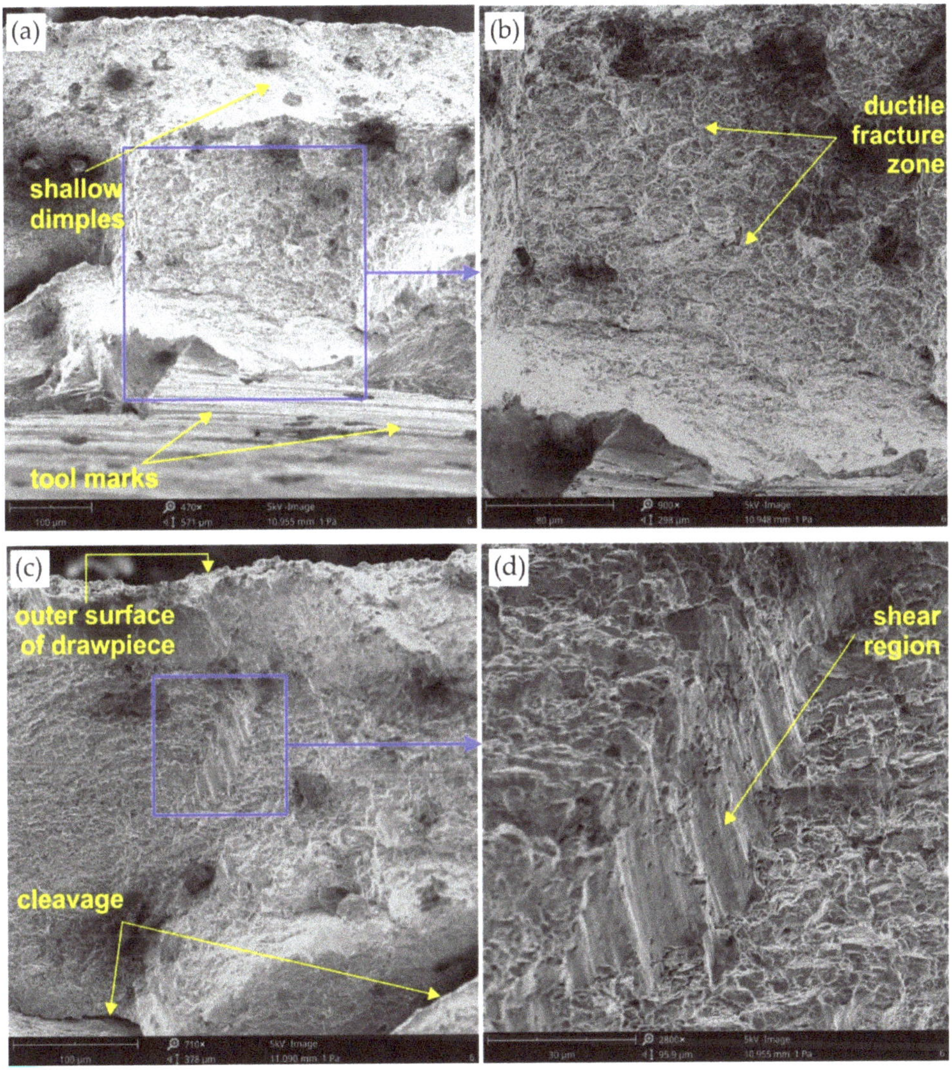

Figure 9. SEM micrographs of the fracture surface of VWACF (maximum formable wall angle $\alpha = 45°$): (**a**) cross-section of the fractured surface, (**b**) view of the near-edge inner surface of the drawpiece, (**c**,**d**) magnification of the ductile fracture area.

The total force F and components of the total force F_{xy} and F_z for the VWACF with the highest wall angle obtained, $\alpha = 66°$, are shown in Figure 10. As can be observed, the values of the total force components vary cyclically. The initial stages of the forming process show a very fast increase in the axial force value. At the same time, the in-plane force value was several times smaller. As the downward movement of the tool increased, the value of the drawpiece radius decreased and the tool began to affect the inner surface of the drawpiece with a more and more lateral surface. Meanwhile, at the beginning of the forming process, the lower surface of the tool tip was the most heavily loaded and caused a rapid increase in axial force.

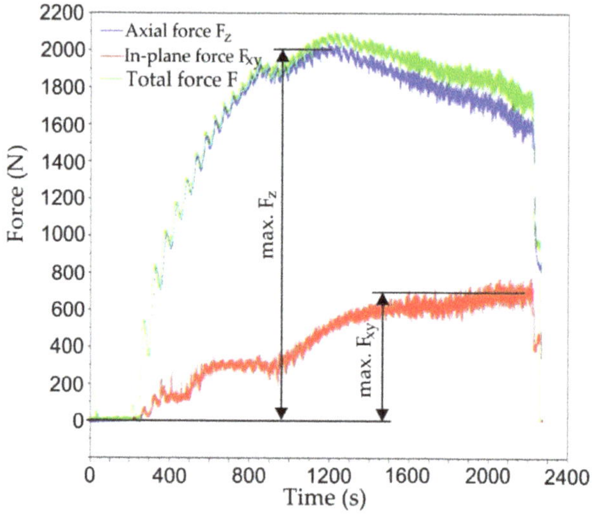

Figure 10. Total force F and components of total force F_{xy} and F_z for the VWACF produced with run 15.

3.2. Maximum Formable Wall Angle

To analyse the effects of the input control factors on the maximum formable wall angle, a model of the response was developed. The adequacy of the RSM models that were built was checked through REML and the coefficient of determination R^2. REML analysis is based on an iterative procedure for estimating the effects of constant and variance components. This procedure begins by defining an initial set of sample parameters, which then are used to estimate the next set of parameters, to replace the set of initial values. The procedure is iterative until convergence is achieved. Then the values of the parameters obtained in the next step differ so little from those obtained in the previous step in such a way that the amount of this difference does not exceed the value of the convergence criterion adopted [42]. The effects of tool rotational speed, feed rate, step size, oil pressure and direction of tool rotation were analysed based on different plots.

Table 5 shows the results of REML analysis for maximum formable wall angle. In REML, whole plots are further divided into smaller portions known as subplots, to which subplot factors were applied [37]. The subplot F-value of 35.53 implied the model is significant. Statistically insignificant factors that affect the process were above the p-value of 0.1. A p-value of 0.05–0.1 indicates marginally significant factors, and a p-value below 0.05 indicates that the factor is significant in the process. Feed rate and tool rotational speed are key parameters that affect the maximum formable wall angle. This is in line with the research findings by Durante et al. [43]. A faster tool rotational speed improves the sheet formability [44]. As Kumar et al. [44] concluded, the formability increase is due to both a positive reduction of the friction effects at the tool sheet interface and localised heating of the sheet material. Although oil pressure is a statistically insignificant factor in the whole model according to the variance components that were estimated using REML the interaction between oil pressure and both tool rotational speed and direction of tool rotation in the subplot is significant. Therefore, to maintain hierarchy, oil pressure was added in the final model.

Table 5. REML analysis of the maximum formable wall angle.

Source	Term df	Error df	F-Value	p-Value	Significance
Whole plot	1	2.99	0.4910	0.5341	not significant
a—oil pressure	1	2.99	0.4910	0.5341	–
Subplot	13	7.00	35.53	<0.0001	significant
B—tool rotational speed	1	7.01	126.43	<0.0001	–
C—feed rate	1	7.00	83.76	<0.0001	–
D—step size	1	7.02	20.44	0.0027	–
E—direction of rotation	1	7.00	58.36	0.0001	–
aB	1	7.00	21.41	0.0024	–
aE	1	7.00	26.87	0.0013	–
BC	1	7.01	37.26	0.0005	–
BE	1	7.00	28.87	0.0010	–
CD	1	7.00	5.95	0.0448	–
CE	1	7.01	28.12	0.0011	–
DE	1	7.00	29.31	0.0010	–
B^2	1	7.00	25.72	0.0014	-
D^2	1	7.00	5.49	0.0516	-

Table 6 presents the restricted maximum likelihood analysis results of the maximum formable wall angle at a confidence interval of 95%. The capability of the REML model is higher than 0.98 indicating that this model fits well with the experimental data. Moreover, the adjusted R^2 value of 0.9957 was in reasonable agreement with the R^2 value of 0.9852. The final equation in terms of coded factors is as follows:

$$\alpha_{max} = 51.27 + 3.37a + 4.1B - 3.14C + 1.67D + 2.06E + 2.46aB + 2.04aE - 2.64BC + 1.83BE + 1.02CD - 1.83CE + 1.83DE + 3.14B^2 - 1.44D^2 \quad (3)$$

Table 6. Fit statistics for REML analysis of the maximum formable wall angle.

Standard Deviation	Mean	Coefficient of Determination R^2	Adjusted Coefficient of Determination R^2	Coefficient of Variance, %
7.40	51.88	0.9852	0.9957	14.27

The function that describes maximum formable wall angle (α_{max}) is given in Equations (4) and (5) in terms of actual factors:

- clockwise direction of tool rotation:

$$\alpha_{max} = 44.95474 - 0.675055a - 0.008857B + 0.005483C + 26.08930D + 0.003382aB - 0.000013BC - 0.002245CD + 0.000021B^2 - 54.62115D^2 \quad (4)$$

- counterclockwise direction of tool rotation:

$$\alpha_{max} = 36.14228 + 1.74114a - 0.002936B + 0.005024C + 38.08526D + 0.003382aB - 0.000013BC - 0.002245CD + 0.000021B^2 - 54.62115D^2 \quad (5)$$

A comparison of the experimental values of the maximum formable wall angle with the values predicted by the REML model is presented in Figure 11. The strong correlation between the predicted and actual values is confirmed by a proportional, close distribution of points along the regression line. The distribution of externally studentized residuals along the horizontal line (Figure 12a,b) shows that the distribution of residuals in the model is normal. The normal distribution of the residuals is necessary to verify the significance of the parameters obtained.

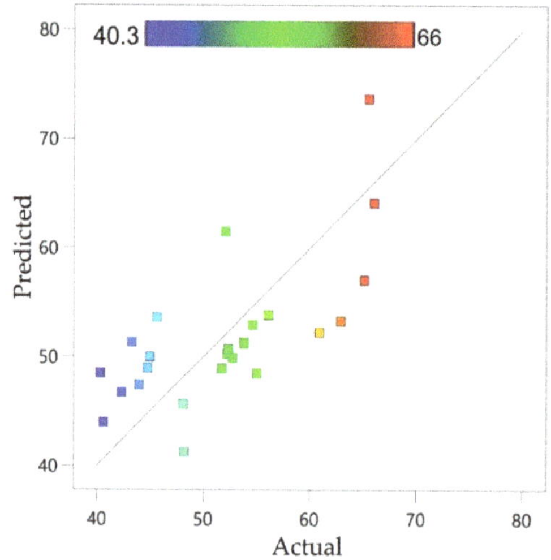

Figure 11. Predicted versus actual response for maximum formable wall angle (units in °).

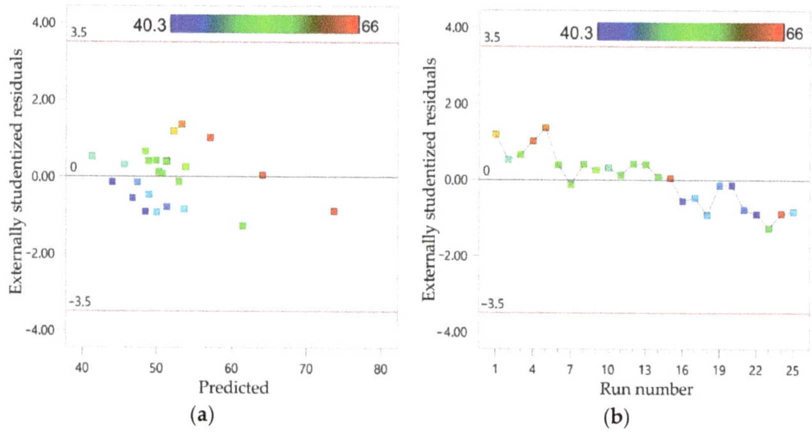

Figure 12. Externally studentized residuals with reference to: (**a**) predicted and (**b**) run number for maximum formable wall angle.

In general, an increase in oil pressure and simultaneously in tool rotational speed increases the maximum formable wall angle (Figure 13a,b). In the case of the anticlockwise direction of tool rotation, larger maximum formable wall angles are obtainable (Figure 13b). This is due to the fact that a higher oil pressure increases the pressure of the sheet against the tool, thus increasing the actual contact surface. Thus, the increased contact area causes more intensive heat generation.

Under these conditions, the high rotational speed of the tool is able to heat the material more and increase its formability. Increasing the tool rotational speed for a given level of feed rate increases the maximum achievable forming angle (Figure 14a,b). The values of maximum formable wall angle during SPIF with a counterclockwise direction of tool rotation (Figure 15b) for the same value of feed rate and step size are greater than during SPIF with a clockwise direction of tool rotation (Figure 15a). A decrease in feed rate leads to an increase in maximum formable wall angle.

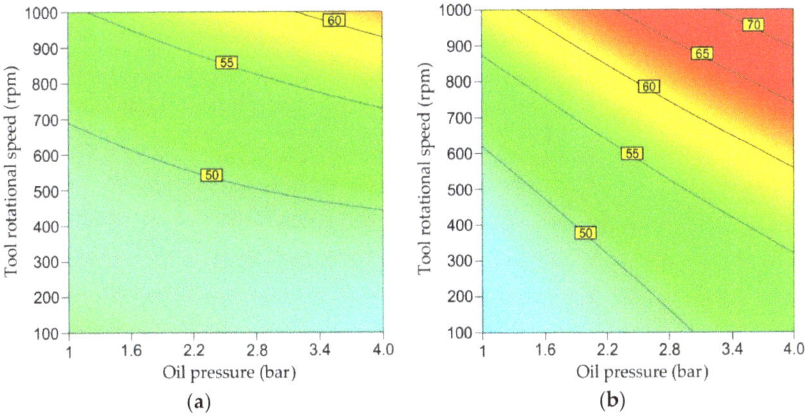

Figure 13. Plots showing the effect of oil pressure and tool rotational speed on maximum formable wall angle α_{max} for (**a**) clockwise and (**b**) anticlockwise direction of tool rotation.

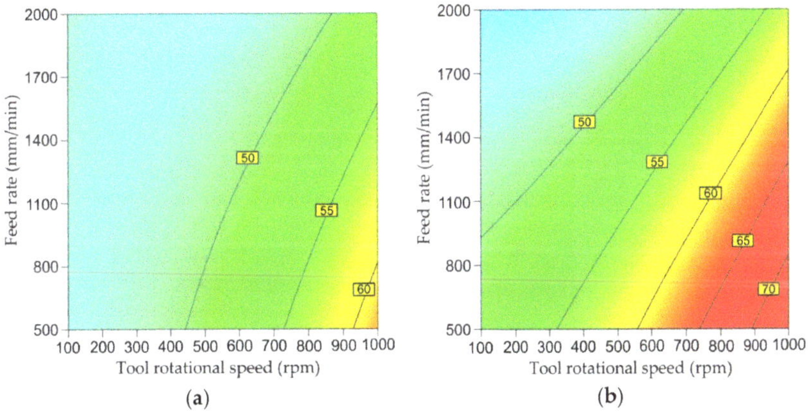

Figure 14. Plots showing the effect of tool rotational speed and feed rate on maximum formable wall angle α_{max} for (**a**) clockwise and (**b**) anticlockwise direction of tool rotation.

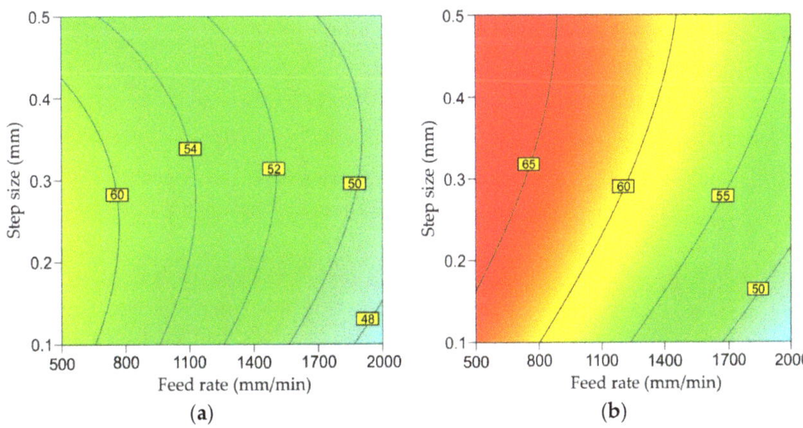

Figure 15. Plots showing the effect of feed rate and step size on maximum formable wall angle α_{max} for (**a**) clockwise and (**b**) anticlockwise direction of tool rotation.

3.3. *In-Plane Force* F_{xy}

The results of REML analysis for in-plane force Fxy are shown in Table 7. The subplot F-value of 28.31 implied that the model was significant. Step size is the most significant factor that affects the in-plane force in SPIF. To complete the part, the tool needs to travel continuously inside the profile, and this required more pushing of the material during forming. The reaction of the material will be greater with high step size values, finally increasing both the in-plane and axial forces [45]. Tool rotational speed and the direction of tool rotation have a significant effect; however, this is slightly smaller than step size. Among all the parameters, the analysis shows that feed rate has the least influence on the in-plane force value. Direction of tool rotation is a statistically insignificant factor in the whole model according to the variance components that were estimated using REML.

Table 7. REML analysis of the in-plane force.

Source	Term df	Error df	F-Value	*p*-Value	Significance
Subplot	5	15.12	28.31	<0.0001	significant
B–tool rotational speed	1	15.12	9.67	0.0071	–
C–feed rate	1	15.08	8.01	0.0126	–
D–step size	1	15.27	105.75	<0.0001	–
E–direction of rotation	1	15.04	9.63	0.0073	–
C^2	1	15.13	11.56	0.0039	-

Table 8 presents the statistics of restricted maximum likelihood analysis of the in-plane force at a confidence interval of 95%. The capability of the REML model is higher than 0.90, indicating that this model fits well with the experimental data. Moreover, the adjusted R^2 value of 0.8614 was in reasonable agreement with the predicted R^2 value.

Table 8. Fit statistics of REML analysis of the in-plane force.

Standard Deviation	Mean	Coefficient of Determination R^2	Adjusted Coefficient of Determination R^2	Coefficient of Variance, %
169.13	578.28	0.9018	0.8614	29.25

The final equation in terms of coded factors is as follows:

$$F_{xy} = 655.44 - 62.87B + 54.95C + 217.94D + 48.24E - 128.24C^2 \quad (6)$$

The function that describes the in-plane force (F_{xy}) is given in Equations (7) and (8) in terms of actual factors:

- clockwise direction of tool rotation:

$$F_{xy} = -208.51375 - 0.178402B + 1.01061C + 935.29813D - 0.000361C^2 \quad (7)$$

- counterclockwise direction of tool rotation:

$$F_{xy} = -146.07492 - 0.178402B + 1.01061C + 935.29813D - 0.000361C^2 \quad (8)$$

Figure 16 shows a comparison of the experimental values of the in-plane force with the values predicted by the REML model. The in-plane force values are evenly distributed around the diagonal. Similarly, externally studentized residuals are proportionally distributed throughout the range of predicted values of in-plane force (Figure 17a). A studentized residual is the quotient resulting from the division of a residual by an estimate of its standard deviation. The normal distribution of the externally studentized residuals is proved by Figure 17b: the residuals lie close to a straight line.

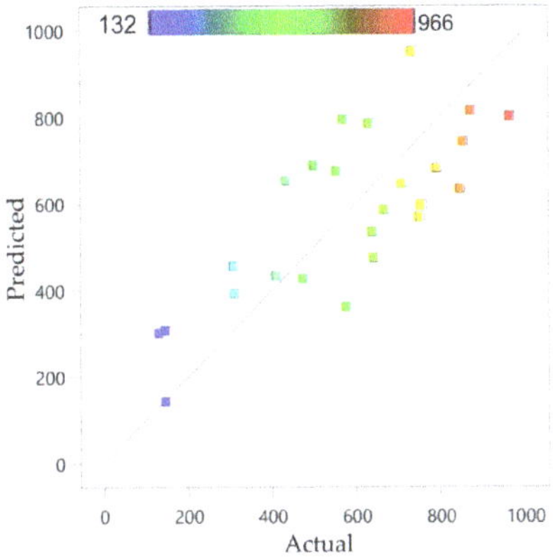

Figure 16. Predicted versus actual response for in-plane force F_{xy} (units in N).

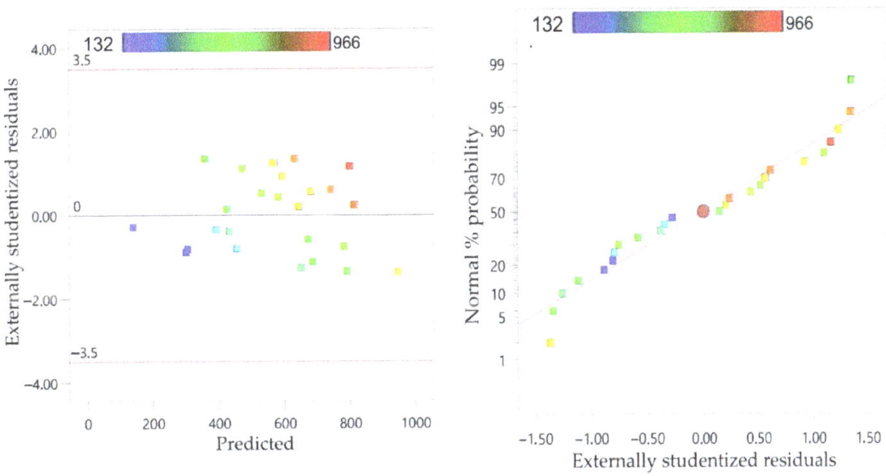

(**a**) (**b**)

Figure 17. (**a**) Externally studentized residuals with reference to predicted F_{xy} values; and (**b**) normal % probability plot of externally studentized residuals.

Reducing the rotational speed of the tool for both directions of tool rotation analysed causes an increase in the in-plane force (Figure 18a,b). The highest in-plane force values are predicted by the REML model in the middle range of feed rate changes and for the lowest tool rotational speed. This phenomenon is observed for both directions of tool rotation. However, in the case of the anticlockwise direction (Figure 18b), the predicted in-plane forces are higher than that for the clockwise direction (Figure 18a).

A very similar trend of the isolines of values of in-plane force is observed after increasing the step size to 0.5 mm (Figure 19a,b). The maximum value of in-plane force is predicted for a feed rate of around 1400 rpm. This conclusion is in line with the results of Baharudin et al. [45]. In the case of the anticlockwise direction of tool rotation (Figure 18b), the area of high in-plane force values extends to the tool rotation range of 100–700 rpm. This proves the increased resistance to movement of the anticlockwise rotating tool. Under these conditions, the tool rotates in the opposite direction to the tool feed, intensifying the frictional interaction of the tool tip with the inner surface of the drawpiece.

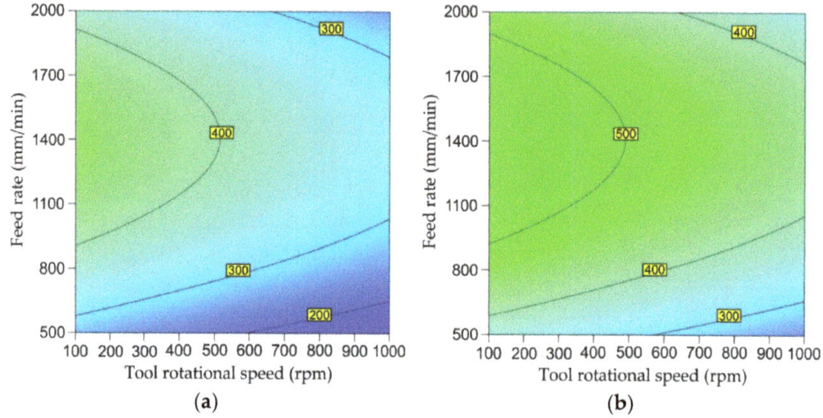

Figure 18. Plots showing the effect of the tool rotational speed and feed rate on the in-plane force F_{xy} for the (**a**) clockwise and (**b**) anticlockwise direction of tool rotation; step size 0.1 mm.

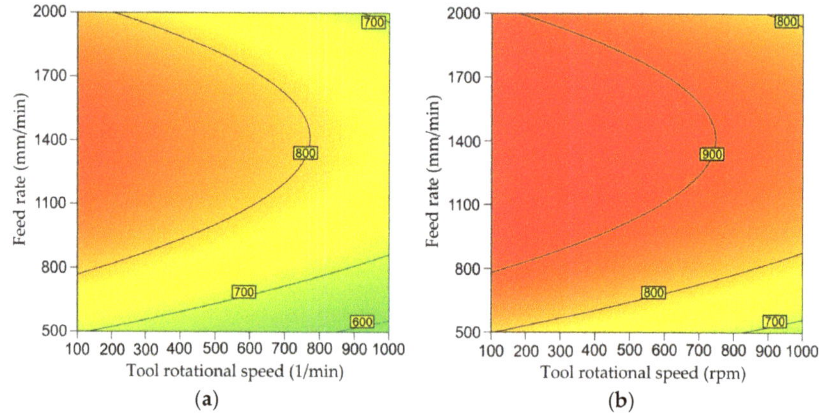

Figure 19. Plots showing the effect of the tool rotational speed and feed rate on in-plane force F_{xy} for (**a**) clockwise and (**b**) anticlockwise direction of tool rotation; step size 0.5 mm.

3.4. Axial Force F_z

Table 9 shows the results of the REML analysis for the axial force F_z. The subplot F-value of 44.58 implied that the model was significant. Tool rotational speed and step size are key parameters that affect the axial force, followed by feed rate. The results of the investigations of Baharudin et al. [45] also showed that the rotation spindle speed was the most dominant parameter affecting the forming forces, followed by feed rate. The interaction between oil pressure and step size in the subplot is significant. Therefore, to maintain hierarchy, oil pressure was included in the final REML model. Direction of tool rotation, as an insignificant parameter, was excluded from the subplot.

Table 10 presents the statistics of REML analysis results of the axial force at a confidence interval of 95%. The capability of the REML model is higher than 0.94, indicating that this model fits well with the experimental data. Moreover, the adjusted R^2 value of 0.9172 was in reasonable agreement with the R^2 value of 0.9483. From the REML results for the responses, it was inferred that the statistical model developed was adequate. Therefore, this model can be used to predict axial force in the design space.

Table 9. REML analysis of the axial force F_Z.

Source	Term df	Error df	F-Value	*p*-Value	Significance
Whole-plot	1	2.52	0.0327	0.8700	not significant
a-oil pressure	1	2.52	0.0327	0.8700	–
Subplot	6	15.10	44.58	<0.0001	significant
B–tool rotational speed	1	15.56	42.77	<0.0001	–
C–feed rate	1	15.11	13.87	0.0020	–
D–step size	1	16.74	169.63	<0.0001	–
aD	1	14.66	7.77	0.0140	–
BC	1	15.58	4.37	0.0535	–
D^2	1	14.17	6.05	0.0273	–

Table 10. Fit statistics of the REML analysis of the axial force.

Standard Deviation	Mean	Coefficient of Determination R^2	Adjusted Coefficient of Determination R^2	Coefficient of Variance, %
113.06	1964.72	0.9483	0.9172	5.75

The final equation in terms of coded factors is as follows:

$$F_Z = 2035.37 + 7.28a - 194.57B + 103.26C + 403.61D - 115.13aD + 74.46BC - 122.30D^2 \quad (9)$$

The function that describes axial force (F_z) is given in Equation (10) in terms of actual factors:

$$F_Z = 1060.64434 + 117.61102a - 0.675077B + 0.027296C + 4827.63140D - 375.50316aD + 0.000199BC - 3146.10137D^2 \quad (10)$$

A comparison of the experimental values of the axial force with the values predicted by the REML model is presented in Figure 20a. The high correlation between the predicted and actual values is confirmed by a proportional, close distribution of points along the regression line. The distribution of externally studentized residuals along the horizontal line (Figure 20b) shows that the distribution of residuals in the model is normal. The normal distribution of the residuals is necessary to verify the significance of the parameters obtained.

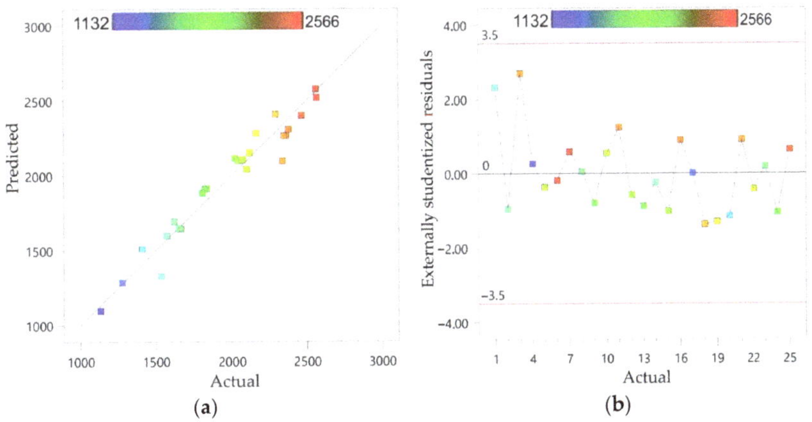

Figure 20. (**a**) Predicted versus actual response for axial forming force (units in N) and (**b**) externally studentized residuals with reference to actual response.

The change in oil pressure at a constant step size value did not have a significant effect on the axial force (Figure 21a,b). However, at the lowest tool rotational speed analysed of 100 rpm with an increase in step size, the axial force increased at a faster rate (Figure 21a) than during forming with the highest tool rotational speed analysed of 1000 rpm (Figure 21b). In general, the smallest axial forces resulting from the pressure of the sheet against the tool occur for the smallest pressure and the smallest step size. The high concentration of isolines in Figure 22a,b proves the strong influence of tool rotational speed and feed rate on the axial force. The interactional effects of these two parameters are inversely proportional. The greatest axial force predicted by the REML model occurs when forming with the lowest tool rotation speed and at the same time with the highest feed rate. SPIF with high tool rotational speed and at the same time low feed rate led to a low axial force.

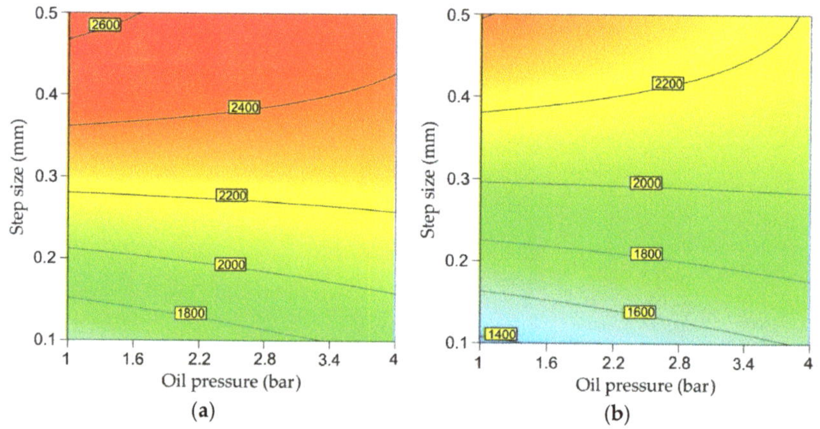

Figure 21. Plots showing the effect of oil pressure and step sizes on axial force F_z for tool rotational speed of (**a**) 100 rpm and (**b**) 1000 rpm.

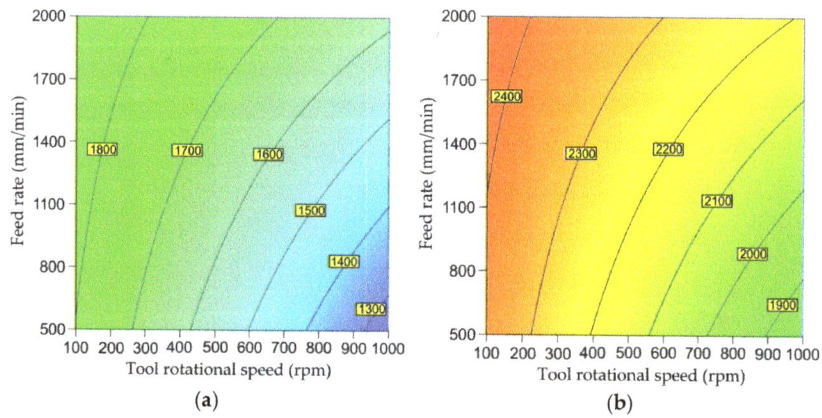

Figure 22. Plots showing the effect of tool rotational speed and feed rate on axial force F_z for step sizes of (**a**) 0.1 mm and (**b**) 0.5 mm.

3.5. Numerical Optimisation

Desirability-based optimisation of the forming parameters was performed based on the desirability of multiple responses [20,44]. The optimisation procedure combines individual desirabilities into a single number and then searches for the greatest overall desirability [20].

Maximum forming angle is an output response that needs to be maximised, and therefore the higher the better is selected as an optimisation constraint. Limitations and constraints for the optimisation of the maximum formable wall angle are shown in Table 11. As an optimal value, the solution with the highest desirability value is chosen.

Table 11. Limits used and goals for optimisation ("+" – clockwise direction, "–"– counterclockwise direction of tool rotation).

Constraint Name	Goal	Lower Limit	Upper Limit
a—oil pressure, bar	is in range	1	4
B—tool rotational speed, rpm	is in range	100	1000
C—feed rate, mm/min	maximise	500	2000
D—step size, mm	maximise	0.1	0.5
E—direction of tool rotation	is in range	"–"	"+"
Maximum formable wall angle α, °	maximise	30	90
Axial force F_z, N	minimise	1132	2566
In-plane force F_{xy}, N	minimise	132	966

Optimisation outcomes are shown in Figure 23. The higher up the ramp, the better the desirability [20]. Considering maximum formable wall angle as output response, the best optimised value of this parameter is 64.19°, which can be obtained when formed with 1000 rpm spindle speed, 2000 mm/min feed rate, 0.38 mm step size, 4 bar oil pressure, and counterclockwise direction of tool rotation.

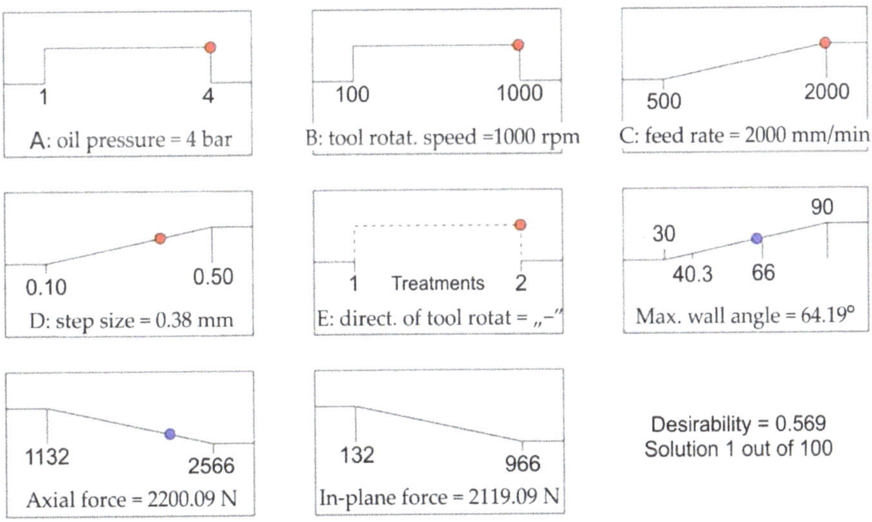

Figure 23. Ramp plot for optimal responses.

The synergistic effect of high feed rate and tool rotational speed result in the most intensive friction stir rotation-assisted heating in SPIF. A high value of oil pressure presses the sheet against the tool tip, and as it is commonly known, greater pressure causes greater friction force. The temperature in the contact zone is the basic parameter that determines the possibility of forming α + β Ti-6Al-4V titanium alloy. Moreover, the toolpath climb strategy, in which the tool and the toolpath move in opposite directions, was seen to provide higher formability in terms of a maximum forming wall angle. This strategy, similar to climb milling, provides a more intense friction interaction between the sheet and the tool than the conventional SPIF strategy.

4. Conclusions

In this paper, a split-plot I-optimal design was used to determine the input parameters for the warm SPIF process on Ti-6Al-4V sheets in order to maximise the formable wall angle and to predict the in-plane and axial forming force components. Models were fitted using REstricted Maximum Likelihood (REML), and p-values are derived using the Kenward–Roger approximation. The following main conclusions can be drawn from the research:

- Observation of the fracture surface of Ti-6Al-4V drawpieces showed that the destruction is the result of a ductile fracture mode. An underside view of the fracture surface, related to the inner surface of the sheet, shows dimples similar to the ones due to ductile fracture under tensile stress. The outer part of the fracture surface presents small, partially formed dimples typical of shear load conditions.
- A simultaneous increase in oil pressure and tool rotational speed increases the maximum formable wall angle. In the case of the anticlockwise direction of tool rotation, larger maximum formable wall angles are obtainable.
- Feed rate and tool rotational speed are key parameters that affect the maximum formable wall angle.
- The high tool rotational speed is able to heat the material more and increase its formability. Increasing the tool rotational speed for a given level of feed rate increases the maximum achievable forming angle.
- Step size is the most significant factor that affects the in-plane SPIF force. Tool rotational speed and the direction of tool rotation have a significant effect; however, it is slightly smaller than step size.
- Tool rotational speed and step size are the most significant factors that affect the axial force, followed by feed rate.
- The change in oil pressure at a constant step size value did not have a significant effect on the axial force.

Future research should determine the influence of process parameters, in particular oil pressure and direction of tool rotation, on the shape and dimensional accuracy of the drawpieces. In addition, the effect of process parameters on the surface roughness parameters (Ra and Rz) of the inner and outer surfaces of drawpieces will be examined. Future studies, including the evaluation of residual stress gradients during both climbing and conventional SPIF, are therefore necessary to obtain a better understanding of the processes and behaviours affecting warm forming of Ti-6Al-4V Ti-based alloy sheets.

Author Contributions: Conceptualization, T.T. and M.S.; methodology, T.T., M.S. and R.O.; validation, M.S. and R.O.; formal analysis, T.T., M.S. and R.O.; investigation, T.T., M.S. and R.O.; data curation, T.T. and M.S.; writing—original draft preparation, T.T.; writing—review and editing, T.T. All authors have read and agreed to the published version of the manuscript.

Funding: This research received no external funding.

Institutional Review Board Statement: Not applicable.

Informed Consent Statement: Not applicable.

Data Availability Statement: The data presented in this study are available on request from the corresponding author.

Conflicts of Interest: The authors declare no conflict of interest.

References

1. Iseki, H.; Naganawa, T.; Kato, K. Flexible and Incremental Sheet Metal Bulging Using a Path-Controlled Spherical Roller: 2nd Report, Vertical Wall Surface Forming of Rectangular Shell Using Multistage Incremental Forming with Spherical and Cylindrical Rollers. *Trans. Jpn. Soc. Mech. Eng. Ser. C* **1996**, *62*, 1600–1605. [CrossRef]
2. Milutinović, M.; Lendjel, R.; Baloš, S.; Zlatanović, D.L.; Sevšek, L.; Pepelnjak, T. Characterisation of Geometrical and Physical Properties of a Stainless Steel Denture Framework Manufactured by Single-Point Incremental Forming. *J. Mater. Res. Technol.* **2021**, *10*, 605–623. [CrossRef]

3. Behera, A.K.; de Sousa, R.A.; Ingarao, G.; Oleksik, V. Single Point Incremental Forming: An Assessment of the Progress and Technology Trends from 2005 to 2015. *J. Manuf. Process.* **2017**, *27*, 37–62. [CrossRef]
4. Żaba, K.; Głodzik, M.; Puchlerska, S.; Pociecha, D.; Nowosielski, M.; Kwiatkowski, M. Analysis of the formability in the incremental sheet forming process. In Proceedings of the 24th International Conference on Metallurgy and Materials-Metal, Brno, Czech Republic, 3–5 June 2015; pp. 1–6.
5. Petek, A.; Gantar, G.; Pepelnjak, T.; Kuzman, K. Economical and Ecological Aspects of Single Point Incremental Forming versus Deep Drawing Technology. *Key Eng. Mater.* **2007**, *344*, 931–938. [CrossRef]
6. Leonhardt, A.; Kurz, G.; Victoria-Hernández, J.; Kräusel, V.; Landgrebe, D.; Letzig, D. Experimental Study on Incremental Sheet Forming of Magnesium Alloy AZ31 with Hot Air Heating. *Procedia Manuf.* **2018**, *15*, 1192–1199. [CrossRef]
7. Ortiz, M.; Penalva, M.; Iriondo, E.; de Lacalle, L.N.L. Accuracy and Surface Quality Improvements in the Manufacturing of Ti-6Al-4V Parts Using Hot Single Point Incremental Forming. *Metals* **2019**, *9*, 697. [CrossRef]
8. Honarpisheh, M.; Abdolhoseini, M.J.; Amini, S. Experimental and Numerical Investigation of the Hot Incremental Forming of Ti-6Al-4V Sheet Using Electrical Current. *Int. J. Adv. Manuf. Technol.* **2016**, *83*, 2027–2037. [CrossRef]
9. Vahdani, M.; Mirnia, M.J.; Bakhshi-Jooybari, M.; Gorji, H. Electric Hot Incremental Sheet Forming of Ti-6Al-4V Titanium, AA6061 Aluminum, and DC01 Steel Sheets. *Int. J. Adv. Manuf. Technol.* **2019**, *103*, 1199–1209. [CrossRef]
10. Göttmann, A.; Diettrich, J.; Bergweiler, G.; Bambach, M.; Hirt, G.; Loosen, P.; Poprawe, R. Laser-Assisted Asymmetric Incremental Sheet Forming of Titanium Sheet Metal Parts. *Prod. Eng.* **2011**, *5*, 263–271. [CrossRef]
11. Göttmann, A.; Bailly, D.; Bergweiler, G.; Bambach, M.; Stollenwerk, J.; Hirt, G.; Loosen, P. A Novel Approach for Temperature Control in ISF Supported by Laser and Resistance Heating. *Int. J. Adv. Manuf. Technol.* **2013**, *67*, 2195–2205. [CrossRef]
12. Grün, P.A.; Uheida, E.H.; Lachmann, L.; Dimitrov, D.; Oosthuizen, G.A. Formability of Titanium Alloy Sheets by Friction Stir Incremental Forming. *Int. J. Adv. Manuf. Technol.* **2018**, *99*, 1993–2003. [CrossRef]
13. Khan, M.S.; Coenen, F.; Dixon, C.; El-Salhi, S.; Penalva, M.; Rivero, A. An Intelligent Process Model: Predicting Springback in Single Point Incremental Forming. *Int. J. Adv. Manuf. Technol.* **2015**, *76*, 2071–2082. [CrossRef]
14. Zhang, Z.; Zhang, H.; Shi, Y.; Moser, N.; Ren, H.; Ehmann, K.F.; Cao, J. Springback Reduction by Annealing for Incremental Sheet Forming. *Procedia Manuf.* **2016**, *5*, 696–706. [CrossRef]
15. Bai, L.; Li, Y.; Yang, M.; Lin, Y.; Yuan, Q.; Zhao, R. Modeling and Analysis of Single Point Incremental Forming Force with Static Pressure Support and Ultrasonic Vibration. *Materials* **2019**, *12*, 1899. [CrossRef] [PubMed]
16. Najm, S.M.; Paniti, I. Study on Effecting Parameters of Flat and Hemispherical End Tools in SPIF of Aluminium Foils. *Teh. Vjesn.* **2020**, *27*, 1844–1849. [CrossRef]
17. Paniti, I.; Viharos, Z.J.; Harangozó, D.; Najm, S.M. Experimental and Numerical Investigation of Single Point Incremental Forming of Aluminium Alloy Foils. *Acta Imeko* **2020**, *9*, 25. [CrossRef]
18. Honarpisheh, M.; Jobedar, M.M.; Alinaghian, I. Multi-Response Optimization on Single-Point Incremental Forming of Hyperbolic Shape Al-1050/Cu Bimetal Using Response Surface Methodology. *Int. J. Adv. Manuf. Technol.* **2018**, *96*, 3069–3080. [CrossRef]
19. Kurra, S.; HR, N.; Regalla, S.; Gupta, A.K. Parametric Study and Multi-Objective Optimization in Single-Point Incremental Forming of Extra Deep Drawing Steel Sheets. *Proc. Inst. Mech. Eng. Part B J. Eng. Manuf.* **2016**, *230*, 825–837. [CrossRef]
20. Szpunar, M.; Ostrowski, R.; Trzepieciński, T.; Kaščák, L'. Central Composite Design Optimisation in Single Point Incremental Forming of Truncated Cones from Commercially Pure Titanium Grade 2 Sheet Metals. *Materials* **2021**, *14*, 3634. [CrossRef] [PubMed]
21. Najm, S.M.; Paniti, I. Artificial Neural Network for Modeling and Investigating the Effects of Forming Tool Characteristics on the Accuracy and Formability of Thin Aluminum Alloy Blanks When Using SPIF. *Int. J. Adv. Manuf. Technol.* **2021**, *114*, 2591–2615. [CrossRef]
22. Maji, K.; Kumar, G. Inverse Analysis and Multi-Objective Optimization of Single-Point Incremental Forming of AA5083 Aluminum Alloy Sheet. *Soft Comput.* **2020**, *24*, 4505–4521. [CrossRef]
23. Durante, M.; Formisano, A.; Langella, A.; Capece Minutolo, F.M. The Influence of Tool Rotation on an Incremental Forming Process. *J. Mater. Process. Technol.* **2009**, *209*, 4621–4626. [CrossRef]
24. Hussain, G.; Gao, L.; Zhang, Z.Y. Formability Evaluation of a Pure Titanium Sheet in the Cold Incremental Forming Process. *Int. J. Adv. Manuf. Technol.* **2008**, *37*, 920–926. [CrossRef]
25. Racz, S.G.; Breaz, R.E.; Tera, M.; Gîrjob, C.; Biriș, C.; Chicea, A.L.; Bologa, O. Incremental Forming of Titanium Ti6Al4V Alloy for Cranioplasty Plates—Decision-Making Process and Technological Approaches. *Metals* **2018**, *8*, 626. [CrossRef]
26. Mohanraj, R.; Elangovan, S. Incremental Sheet Metal Forming of Ti-6Al-4V Alloy for Aerospace Application. *Trans. Can. Soc. Mech. Eng.* **2020**, *44*, 56–64. [CrossRef]
27. Palumbo, G.; Brandizzi, M.; Cervelli, G.; Fracchiolla, M. Investigations about the Single Point Incremental Forming of Anisotropic Titanium Alloy Sheets. *Adv. Mater. Res.* **2011**, *264–265*, 188–193. [CrossRef]
28. Naranjo, J.; Miguel, V.; Martínez, A.; Coello, J.; Manjabacas, M.C.; Valera, J. Influence of Temperature on Alloy Ti6Al4V Formability during the Warm SPIF Process. *Procedia Eng.* **2017**, *207*, 866–871. [CrossRef]
29. Sbayti, M.; Bahloul, R.; Belhadjsalah, H. Numerical Modeling of Hot Incremental Forming Process for Biomedical Application. In *Design and Modeling of Mechanical Systems—III*; Haddar, M., Chaari, F., Benamara, A., Chouchane, M., Karra, C., Aifaoui, N., Eds.; Lecture Notes in Mechanical Engineering; Springer International Publishing: Cham, Switzerland, 2018; pp. 881–891. ISBN 978-3-319-66696-9.

30. Sbayti, M.; Ghiotti, A.; Bahloul, R.; Belhadjsalah, H.; Bruschi, S. Finite Element Analysis of Hot Single Point Incremental Forming of Hip Prostheses. *MATEC Web Conf.* **2016**, *80*, 14006. [CrossRef]
31. Oleksik, V.; Trzepieciński, T.; Szpunar, M.; Chodoła, Ł.; Ficek, D.; Szczęsny, I. Single-Point Incremental Forming of Titanium and Titanium Alloy Sheets. *Materials* **2021**, *14*, 6372. [CrossRef] [PubMed]
32. Trzepieciński, T.; Oleksik, V.; Pepelnjak, T.; Najm, S.M.; Paniti, I.; Maji, K. Emerging Trends in Single Point Incremental Sheet Forming of Lightweight Metals. *Metals* **2021**, *11*, 1188. [CrossRef]
33. Ambrogio, G.; Filice, L.; Manco, G.L. Warm Incremental Forming of Magnesium Alloy AZ31. *CIRP Ann.* **2008**, *57*, 257–260. [CrossRef]
34. Pederson, R. Microstructure and Phase Transformation of Ti-6Al-4V. Licentiate Thesis, Luleå University of Technology, Luleå, Sweden, May 2002.
35. Villa, M.; Brooks, J.W.; Turner, R.P.; Wang, H.; Boitout, F.; Ward, R.M. Microstructural Modeling of the $\alpha + \beta$ Phase in Ti-6Al-4V: A Diffusion-Based Approach. *Metall. Mater. Trans.* **2019**, *50*, 2898–2911. [CrossRef]
36. *ISO 5832-3:2016*; Implants for Surgery—Metallic Materials—Part 3: Wrought Titanium 6-Aluminium 4-Vanadium Alloy. International Organization for Standardization: Geneva, Switzerland, 2016.
37. Jones, B.; Goos, P. I-Optimal Versus D-Optimal Split-Plot Response Surface Designs. *J. Qual. Technol.* **2012**, *44*, 85–101. [CrossRef]
38. Balicki, S. Optymalizacja procesów jednostkowych w technologii organicznej. *Przemysł Chem.* **2021**, *5*, 490–497. [CrossRef]
39. Turnage, S.A.; Darling, K.A.; Rajagopalan, M.; Whittington, W.R.; Tschopp, M.A.; Peralta, P.; Solanki, K.N. Quantifying Structure-Property Relationships during Resistance Spot Welding of an Aluminum 6061-T6 Joint. Available online: https://arxiv.org/ftp/arxiv/papers/1605/1605.04251.pdf (accessed on 4 November 2021).
40. ASM International. *ASM Handbook Volume 12: Fractography*; ASM International: Materials Park, OH, USA, 2001.
41. Valoppi, B.; Zhang, Z.; Deng, M.; Ghiotti, A.; Bruschi, S.; Ehmann, K.F.; Cao, J. On the Fracture Characterization in Double-Sided Incremental Forming of Ti6Al4V Sheets at Elevated Temperatures. *Procedia Manuf.* **2017**, *10*, 407–416. [CrossRef]
42. Radkiewicz, P.; Zieliński, M.W. Hierarchiczne modele liniowe. Co nam dają i kiedy warto je stosować. *Psychol. Społeczna* **2010**, *5*, 217–233.
43. Durante, M.; Formisano, A.; Lambiase, F. Formability of polycarbonate sheets in single-point incremental forming. *Int. J. Adv. Manuf. Technol.* **2016**, *102*, 2049–2062. [CrossRef]
44. Kumar, S.P.; Elangovan, S.; Mohanraj, R. Experimental study on single point incremental forming of Inconel 718. *Trans. Can. Soc. Mech. Eng.* **2020**, *44*, 179–188. [CrossRef]
45. Baharudin, B.T.H.T.; Azpen, Q.M.; Sulaima, S.; Mustapha, F. Experimental Investigation of Forming Forces in Frictional Stir Incremental Forming of Aluminum Alloy AA6061-T6. *Metals* **2017**, *7*, 484. [CrossRef]

metals

MDPI

Article

Research on Shape Control Characteristics of Non-oriented Silicon Steel for UCMW Cold Rolling Mill

Hao Tao [1], Hongbo Li [1], Jian Shao [2,*], Jie Zhang [1], Yujin Liu [3] and Xuechang You [3]

1 School of Mechanical Engineering, University of Science and Technology Beijing, Beijing 100083, China; 17888822456@163.com (H.T.); lihongbo@ustb.edu.cn (H.L.); ZhangJie@ustb.edu.cn (J.Z.)
2 National Engineering Research Center for Advanced Rolling Technology, University of Science and Technology Beijing, Beijing 100083, China
3 Shougang Zhixin Qian'an Electromagnetic Material Co., Ltd., Qian'an 064400, China; liuyujin@sgqg.com (Y.L.); youxuechang@sgqg.com (X.Y.)
* Correspondence: jianshao@ustb.edu.cn; Tel.: +86-10-6233-6320

Received: 13 July 2020; Accepted: 5 August 2020; Published: 7 August 2020

Abstract: In order to analyze the flatness control characteristics for a certain UCMW (Universal Crown Mill with Work roll shifting) cold rolling mill, combined with the actual parameters in the field, a static simulation model of the quarter roll systems of the UCMW cold rolling mill was established by the ANSYS finite element software. The bearing roll gaps under the factors of the unit width rolling force, the roll bending force and the roll shift were calculated, which reflects the shape control characteristics and has a great influence on the friction and lubrication characteristics between the roll gaps. Additionally, the shape control strategy of the process parameters in the field was put forward. The results show that, at first, the work roll shift is the most effective shape control means, while the current-used range of the intermediate roll shift cannot make full use of the roll end contour of the intermediate roll, so the intermediate roll negative shift should be considered for shape control. At second, the excessive rolling force goes against the shape control, so the rolling force of each stand should be reasonably distributed. Finally, the shape control ability of the bending force is relatively weak, so the range of the work roll bending force should be appropriately increased.

Keywords: UCMW cold rolling mill; finite element simulation; roll gap friction; shape control

1. Introduction

In recent years, with the increasing of downstream manufacturers' requirements on the quality of cold-rolled strip products, the shape quality requirements on cold-rolled strips have become more stringent. Therefore, rolling mills with high-precision shape control ability have been more and more widely used. The UCMW (Universal Crown Mill with Work roll shifting) cold rolling mill, developed on the basis of the HC (High Crown) rolling mill, is the most representative one. It has been widely used in rolling non-oriented silicon steel with strict quality requirements because of its great performance in shape control [1–4]. The UCMW cold rolling mill is equipped with a work roll positive and negative bending roll mechanism and an intermediate roll positive bending roll mechanism to enhance the regulating ability of the bearing roll gap crown. Additionally, it has an intermediate roll axial traverse mechanism. The transverse rigidity of the bearing roll gap can be infinitely increased theoretically by adjusting the position of the intermediate roll, so the rolling stability is improved. Moreover, a single taper work roll axial traverse mechanism has been equipped to effectively control the strip edge drop and reduce the strip edge cut loss, so that the control accuracy of the strip lateral thick difference can be improved. With reasonable cooperation of the different shape control means, it can effectively improve the shape control ability [5–10].

During the rolling process, the shape of the rolled product is actually determined by the roll gap. Therefore, the bearing roll gap is an important basis for reflecting the shape control characteristics. The shape control characteristics of the rolling mill can be fully played by researching the influence of different factors on the bearing roll gap [11–14]. For bearing roll gaps, Z. D. Han et al. [15] analyzed the influence of factors such as rolling force, bending force, and strip width on the bearing roll gaps during the strip production process by using the influence function method. X. C. Wang et al. [16] analyzed the influence of the work roll shape height on the regulation ability of the rolling mill edge drop by using the partition matrix iteration method and the influence function method. Aiming at the problem of high precision shape quality of cold-rolled strips, J. G. Cao et al. [17] proposed a control strategy which is a whole-unit integrated shape for the strip cold rolling mill.

In order to further grasp the shape control performance of the UCMW cold rolling mill during the rolling of non-oriented silicon steel, the effects of different factors on the bearing roll gap are grasped comprehensively by the finite element simulation combined with the parameters in the field, which can provide a theoretical basis for solving the shape problem in the field.

2. Establishment of Finite Element Model of UCMW Cold Rolling Mill Roll System

2.1. Rolling Mill Parameters and Model Establishment

Taking the UCMW cold rolling mill of a non-oriented silicon steel rolling production line as the research object, the static simulation model of the quarter roll system of the UCMW cold rolling mill shown in Figure 1 was established by using the general ANSYS finite element software. In the actual rolling process, the bearing roll gap of the rolling mill is influenced by many factors such as strip tension, rolling torque, material characteristics, lubrication condition, rolling parameters and rolling temperature, and these factors always change during the actual rolling process, which always influences the bearing roll gap. It is impossible for the finite element model to take into account all the factors, and the necessary assumptions and simplifications must be made:

(1) Ignore the effects of the strip tension and the rolling torque.
(2) The material of the roll is uniform and isotropic.
(3) The axis of the roll is coplanar, which is symmetrical.
(4) There is no relative sliding between the rollers.
(5) The influence of rolling temperature is not considered, namely, ignore the influence of thermal crown.
(6) Only the deformation of the roll system below the rolling line is calculated, and the interaction of the strip and the roll system is reflected by the uniformly distributed pressure.

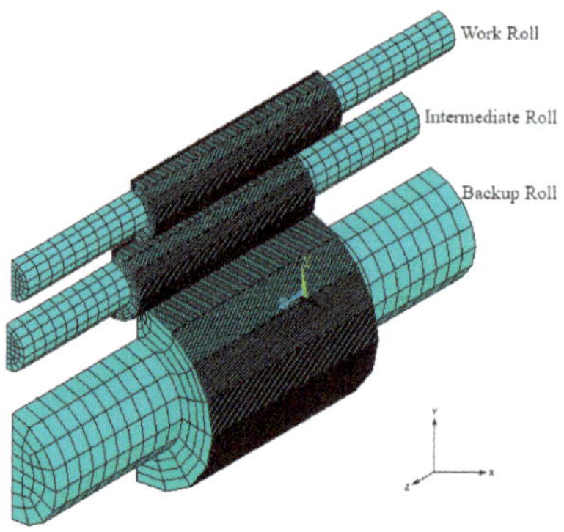

Figure 1. Universal Crown Mill with Work roll shifting (UCMW) roll system finite element model.

The literatures [18–20] show that this model has good applicability in calculating the influence of parameters on the roll gap during rolling. The main geometric parameters of the roll system are shown in Table 1. Among them, the shape of the backup roll is a double-sided taper roll, the shape of the intermediate roll and the work roll are single-sided taper roll, and the roll shape structure of the single-sided taper work roll is shown in Figure 2, taper length L_1 = 150 mm. The roll shape of the work roll and the intermediate roll are input by points, so that the actual situation of the roll shape can be accurately reflected, and the initial end roll shape parameters are shown in Table 2.

Table 1. The main parameters of roll system.

Roll	Roll Neck (Diameter/mm × Length/mm)	Roll Body (Diameter/mm × Length/mm)
Backup roll	770 × 1090	1300 × 1420
Intermediate roll	320 × 975	490 × 1500
Work roll	280 × 1035	425 × 1600

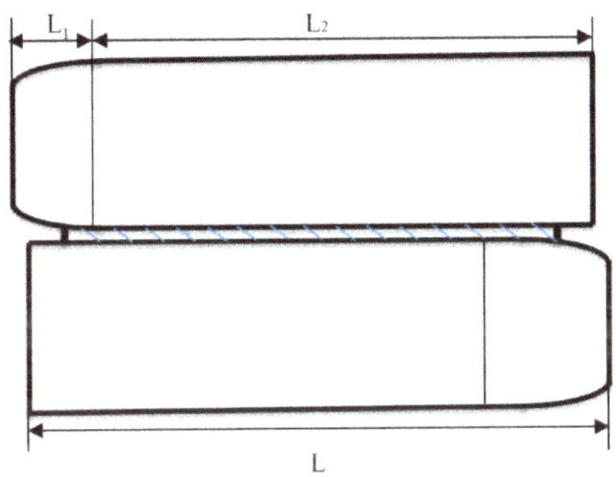

Figure 2. Roll shape structure of single-sided taper work roll.

Table 2. The initial end roll shape parameters.

Roll Shape	Distance from the Midpoint of the Roll/mm	Radius Difference/mm
The work roll	800	0.38
	790	0.33102
	780	0.28542
	770	0.24320
	760	0.20435
	750	0.16889
	740	0.13680
	730	0.10809
	720	0.08276
	710	0.06080
	700	0.04222
	690	0.02702
	680	0.01520
	670	0.00676
	660	0.00147
	650	0
The Intermediate roll	−750	2.4
	−740	1.53473
	−730	0.86273
	−720	0.38326
	−710	0.09579
	−700	0

In order to make the finite element model consistent with the reality, the material characteristics of rolls use actual characteristics (Young's modulus E = 2.1 × 105 MPa, Poisson's ratio v = 0.3). The roll system is divided by the Solid45 element and the contact area between the work roll and strip is subdivided by the Solid95 element. The contact element is added to the surface between the rollers, the backup roll and work roll is Target170 element, and the intermediate roll is Contact174 element. As a result, this model is divided into 38,304 elements and 44,113 nodes.

Based on the actual force conditions of the rolling mill and the characteristics of the rolling process, these displacement constraints are imposed on the established finite element model: (1) Apply symmetry constraints and X direction displacement constraints on all nodes of the YZ plane of the roll system: UX = 0; (2) Apply a Y direction displacement constraint at the midpoint of the top contact line on the backup roll: UY = 0; (3) Apply a Z direction displacement constraint at the geometric centers of the work roll, intermediate roll and backup roll: UZ = 0. The loads are further applied to the model. (1) Rolling force: it is applied on the upper contact line of the work roll as a uniform load. (2) Bending force of the intermediate roll and the work roll: it is applied in the center of the journal section of both ends of the intermediate roll and the work roll as a concentrated force. The bearing condition of the roller system and the bearing roll gap are shown in Figure 3. The deformation of the work roll surface is calculated by finite element simulation, and then the bearing roll gap can be calculated.

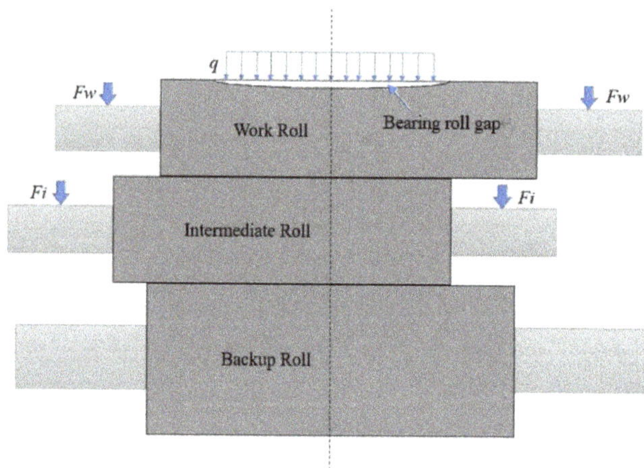

Figure 3. The bearing condition of the roller system.

2.2. Field Data Acquisition and Simulation Condition Design

In order to analyze the shape control characteristics of the UCMW mill, this paper mainly focuses on several main means of shape control, such as: unit width rolling force q, intermediate roll bending force Fi, work roll bending force Fw, intermediate roll shift Si and work roll shift Sw. Among them, the definition of the intermediate roll shift Si and the work roll shift Sw is shown in Figure 4. According to the date tracked and the various parameters of each stand in the field, the simulation condition is designed as shown in Table 3. Other factors are controlled as the initial state when researching the influence of a single factor, setting Fi = 80 kN, Fw = 80 kN, Si = 0 mm, Sw = 0 mm, q = 9 kN/mm as the initial state.

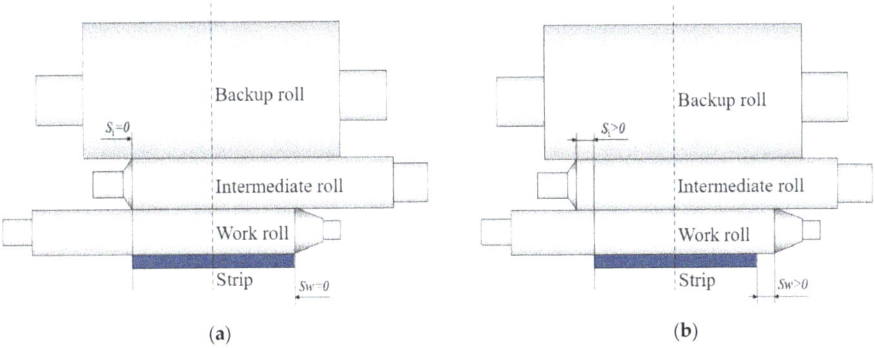

Figure 4. Roll shift amount. (**a**) Zero shifting roll position, (**b**) positive shifting roll position.

Table 3. Simulation condition.

Parameters	On-Site Value Range	Simulation Value
Strip width B/mm	1000~1200	1000, 1100, 1200
Unit width rolling force *q*/(kN/mm)	-	6, 9, 12
Intermediate roll bending force *Fi*/kN	80~220	80, 150, 220
Work roll bending force *Fw*/kN	80~220	80, 150, 220
Intermediate roll shift *Si*/mm	0~80	−40, 0, 40, 80
Work roll shift *Sw*/mm	−80~80	−80, −60, −40, 0, 40, 80

3. Experimental Verification of the Model

In order to verify the accuracy of the model, the lateral thick difference of the roll gap calculated by the model is compared with the lateral thick difference of the strip obtained by field measurement. It is found that the model calculation result is in the same trend as the field measurement result, and the error is also acceptable. One of the working condition parameters is that the strip width B = 1240 mm, the rolling force Q = 11,384.7 kN, the work roll bending force *Fw* = 195 kN, the intermediate roll bending force *Fi* = 228 kN, the work roll shift *Sw* = −70 mm and the intermediate roll shift *Si* = 10 mm. The calculation conditions of the simulation are basically identical with the rolling conditions of the first stand of the non-oriented silicon steel grade 50SW1300 in the field. The comparison between the simulation results and the measured results is shown in Figure 5.

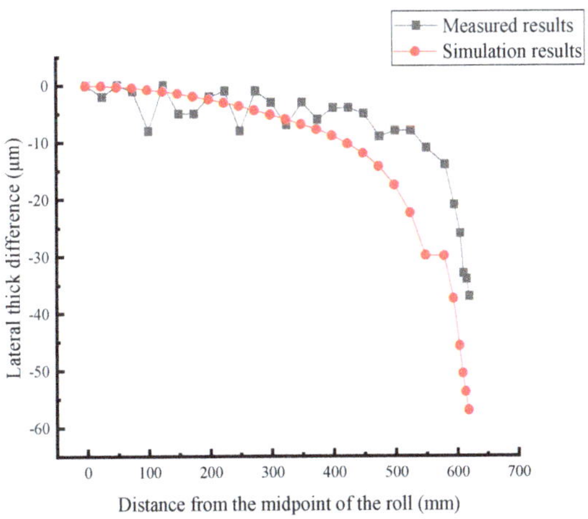

Figure 5. The comparison between the simulation results and the measured results.

It can be seen that although the simulation results are larger than the measured results, whether the crown or the edge drop, the error is also acceptable. The simulation result is larger because this model does not consider some factors of the strip (strip deformation resistance, tension, etc.), so the simulation result is satisfactory. This model is suitable for analyzing the influence of the parameters on the deformation of the roll system.

4. Analysis of Shape Control Characteristics for UCMW Cold Rolling Mill

4.1. Effect of Unit Width Rolling Force on Roll Gap Shape

The effect of the unit width rolling force on roll gap shape is shown in Figure 6. The results of the 1100 mm width strip in Figure 6a show that the bearing roll gap crown decreases from 207.8 μm to 74.6 μm (i.e., 64.1% drop) and the bearing roll gap edge drop decreases from 96.4 μm to 39.7 μm (i.e., 58.8% drop) when the unit width rolling force decreases from 12 kN/mm to 6 kN/mm, which easily causes the shape problem of excessive crown, especially the excessive edge drops for the S1 and S2 stands with the relatively large rolling forces. Then, it can be seen from Figure 6b that at the same unit width rolling force, the roll gap crown changes little with the increase in strip width. As shown in Figure 6c, the variation of the roll gap crown corresponding to the variation of the unit width rolling force is basically the same for the strip with different widths, which indicates that there is little difference between the crown and edge drop control when using the same unit width rolling force to roll the non-oriented silicon steel with different widths.

4.2. Effect of Intermediate Roll Bending Force on Roll Gap Shape

The effect of intermediate roll bending force on roll gap shape is shown in Figure 7. The analysis results of the 1100 mm width strip in Figure 7a show that the bearing roll gap crown decreases from 141.3 μm to 135.5 μm (i.e., 4.1% drop) and the bearing roll gap edge drop decreases from 68.1 μm to 66.9 μm (i.e., 1.8% drop) when the intermediate roll bending force increases from 80 kN to 220 kN, which is the minimum and maximum values of the actual rolling parameters. As shown in Figure 7b, the variation of the roll gap crown corresponding to the variation of the intermediate roll bending force is basically the same for the strip with different widths, which indicates that the intermediate roll bending force has basically no effect on the shape control within the current-used range of the intermediate roll bending force in the field.

4.3. Effect of Work Roll Bending Force on Roll Gap Shape

The effect of the work roll bending force on roll gap shape is shown in Figure 8. The results of the 1100 mm width strip in Figure 8a show that the bearing roll gap crown decreases from 141.3 μm to 80 μm (i.e., 43.4% drop) and the bearing roll gap edge drop decreases from 68.1 μm to 49.1 μm (i.e., 27.9% drop) when the work roll bending force increases from 80 kN to 220 kN, which are the minimum and maximum values of the actual rolling parameters. It indicates that the shape control ability of the UCMW rolling mill is still not strong within the current-used range of the roll bending force in the field, but the shape control ability of the work roll bending force is more than 10 times that of the intermediate roll bending force within the same range of roll bending force. As shown in Figure 8b, the variation of the roll gap crown corresponding to the variation of the work roll bending force is basically the same for the strip with different widths.

It can be seen that the work roll bending force should be the main control means based on the results of the effects of the intermediate roll bending force and the work roll bending force on the roll gap shape. More importantly, the current-used range of the work roll bending force in the field should be increased to better exert the shape control characteristics of the bending force.

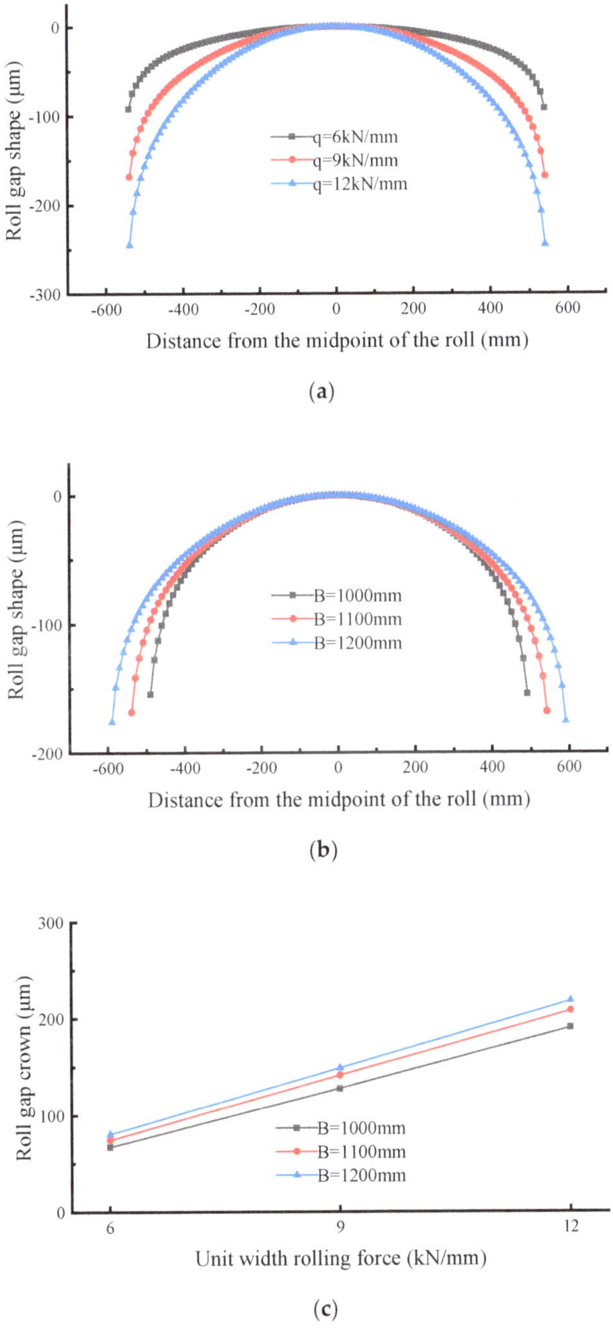

Figure 6. Effect of unit width rolling force on roll gap shape. (**a**) Roll gap shape under different unit width rolling forces, (**b**) roll gap shape of strip with different widths under the same unit width rolling force, (**c**) effect of unit width rolling force on roll gap crown.

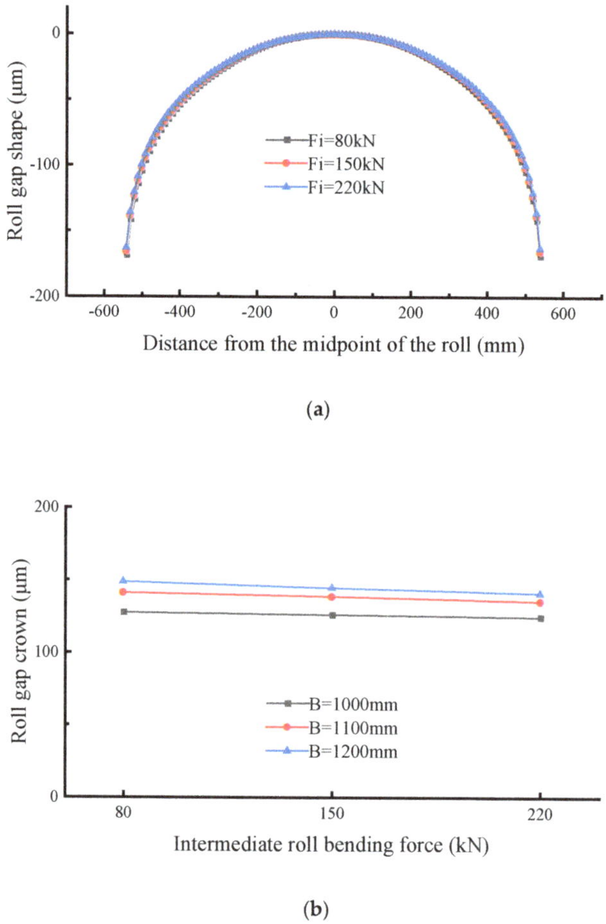

Figure 7. Effect of intermediate roll bending force on roll gap shape. (**a**) Roll gap shape under different intermediate roll bending forces, (**b**) effect of intermediate roll bending force on roll gap crown.

4.4. *Effect of Intermediate Roll Shift on Roll Gap Shape*

The effect of the work roll bending force on roll gap shape is shown in Figure 9. The results of the 1100 mm width strip in Figure 9a show that the bearing roll gap crown decreases from 230.2 μm to 141.3 μm (i.e., 38.6% drop) and the bearing roll gap edge drop decreases from 90.8 μm to 68.1 μm (i.e., 25% drop) when the intermediate roll shift decreases from 80 mm to 0 mm, which is the current-used range of intermediate roll shift in the field. As shown in Figure 9b, the variation of the roll gap crown corresponding to the variation of the intermediate roll shift is slightly different for the strip with different widths; the variation of the 1000 mm width strip is 86.2% of the 1200 mm width strip. The above results show that within the current-used range of the intermediate roll shift in the field, the smaller the roll shift is, the better the shape control ability shows. In fact, the current-used range of the intermediate roll shift cannot make full use of the roll end contour of the intermediate roll. As shown in Figure 9a, the roll gap crown and edge drop are greatly decreased when the intermediate roll is negatively shifted. They decrease by 41.5% and 21.1%, respectively, when the intermediate roll shift decreases from 0mm to −40 mm, which indicates that the intermediate roll negative shift should be considered to make full use of the roll end contour of the intermediate roll to shape control.

4.5. *Effect of Work Roll Shift on Roll Gap Shape*

The effect of the work roll bending force on roll gap shape is shown in Figure 10. The results of the 1100 mm width strip in Figure 10a show that the work roll positive shift (i.e., $Sw > 0$ mm) has little influence on the roll gap crown and edge drop. The decreased ratio of the crown and edge drop between $Sw = 80$ mm and $Sw = 0$ mm is 20.6% and 12.9%. However, the roll gap edge region has a significant change which makes the roll gap crown and edge drop significantly decreased, especially the edge drop, as the work roll negative shift (i.e., $Sw < 0$ mm) increases. The roll gap crown and edge drop decrease by 22.8% and 20.2%, respectively, when the work roll shift decreases from 0 mm to −40 mm, which indicates that the control ability of the work roll shift to the roll gap crown is smaller than the intermediate roll shift, but it is stronger for roll gap edge drop compared with the results of the intermediate roll shift $Si = -40$ mm. The roll gap edge region has a significant local change when the work roll shift is $Sw = -60$ mm and $Sw = -80$ mm. The edge drop value at $Sw = -60$ mm is 44.8% lower than $Sw = 0$ mm. Moreover, the phenomenon of the local increase in the roll gap edge appears when the work roll shift is $Sw = -80$ mm, which indicates that the more the roll end contour enters in the strip width range, the stronger the local control ability of the strip edge drop shows.

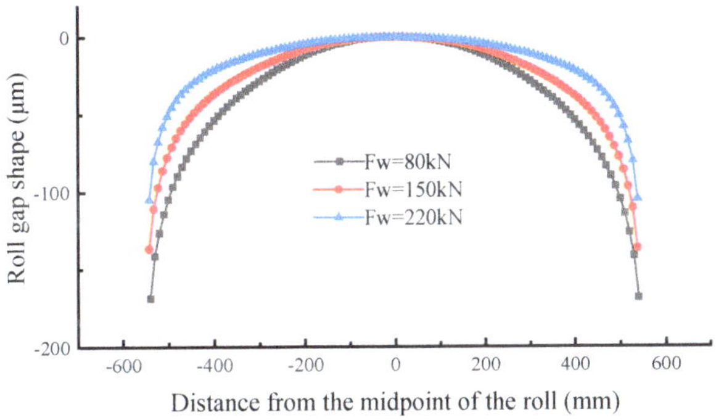

(**a**)

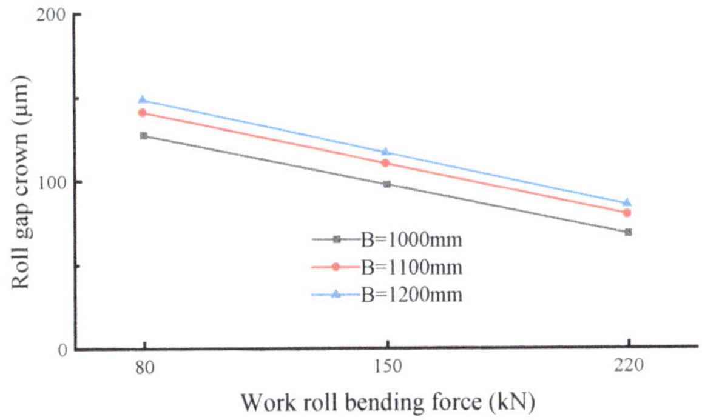

(**b**)

Figure 8. Effect of work roll bending force on roll gap shape. (**a**) Roll gap shape under different intermediate roll bending forces, (**b**) effect of intermediate roll bending force on roll gap crown.

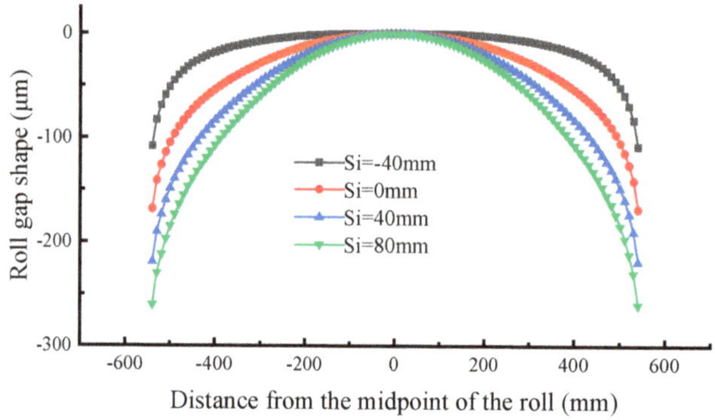

(**a**)

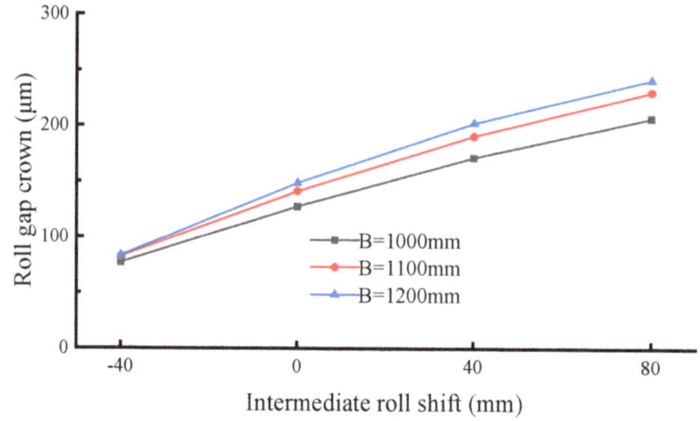

(**b**)

Figure 9. Effect of intermediate roll shift on roll gap shape. (**a**) Roll gap shape under different intermediate roll shifts, (**b**) effect of intermediate roll shift on roll gap crown.

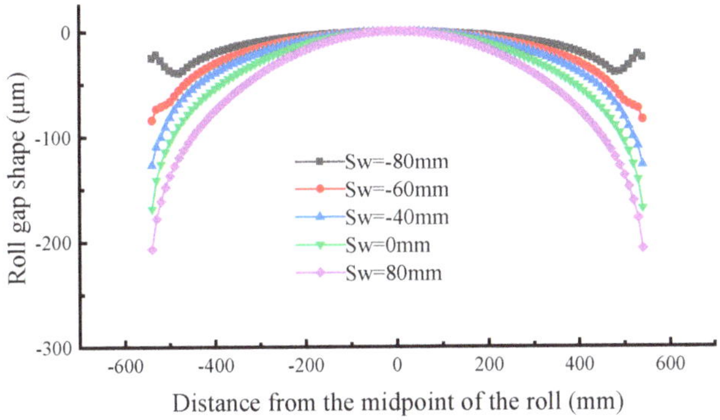

(a)

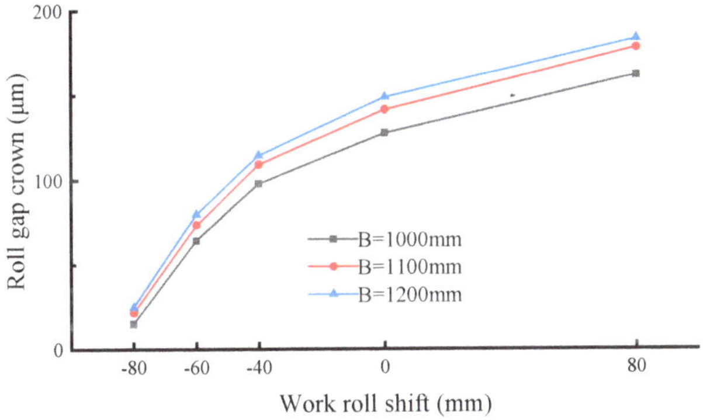

(b)

Figure 10. Effect of work roll shift on roll gap shape. (**a**) Roll gap shape under different work roll shifts, (**b**) effect of work roll shift on roll gap crown.

5. Conclusions

Through the above finite element simulation analysis, the conclusions are:

(1) According to the results of the effect of the unit width rolling force on roll gap shape, it can be seen that the rolling force of the S1 and S2 stands is relatively large in the field, which easily causes the shape problem of excessive crown and edge drop. Meanwhile, the friction condition between the roll and the strip will deteriorate because of the large rolling force. Therefore, the rolling force of each stand needs to be distributed appropriately, with an aim to avoid too much rolling force in a certain stand.

(2) The results of the intermediate roll bending force and the work roll bending force show that the intermediate roll bending force has basically no effect on the shape control. Although the shape control ability of the work roll bending force is more than 10 times that of the intermediate roll bending force, it is still insufficient. Therefore, the range of the work roll bending force should be appropriately increased to fully use the control ability of the bending force.

(3) The smaller the intermediate roll shift is, the better the control ability of the crown and edge drop shows within the current-used range of the intermediate roll shift. Moreover, the means of the

intermediate roll negative shift should be taken to make use of the roll end contour of the intermediate roll for shape control.

(4) The influence of the work roll positive shift on the roll gap crown and edge drop is small. However, the work roll negative shift has significant control on the crown and edge drop because the roll end contour can enter in the strip width range. Notably, the more the roll end contour enters the strip width range, the stronger the local control ability of the strip edge drop shows.

Author Contributions: Conceptualization, H.L., J.S. and J.Z.; Date curating, H.T.; investigation, H.T.; methodology, H.T.; resources, Y.L. and X.Y.; writing—original draft preparation, H.T.; writing—review and editing, H.L., J.S. and J.Z.; project administration, H.L. and J.Z. All authors have read and agreed to the published version of the manuscript.

Funding: This research was funded by the National Key Technology R&D Program of the 12th Five-year Plan of China, grant number 2015BAF30B01. The APC was funded by University of Science and Technology Beijing.

Acknowledgments: First of all, I would like to thank the laboratory teachers and brothers for their help in my study, and they have given me great help in simulation processing. Secondly, thank the Shougang Zhixin Qian'an Electromagnetic Material Co., Ltd. for providing us with cold-rolled non-oriented silicon steel for data testing. Finally, I am very grateful to the three reviewers for their valuable comments, which made this article a great improvement in details.

Conflicts of Interest: The authors declare no conflict of interest.

References

1. Peng, Y.; Liu, H.M.; Wang, D.C. Simulation of type selection for 6-high cold tandem mill based on shape control ability. *J. Cent. South Univ. Technol.* **2007**, *14*, 278–284. [CrossRef]
2. Zhou, X.M.; Cheng, X.X. The Analysis of Influence on the Edge Drop of Tandem Cold Rolling Mill. *Appl. Mech. Mater.* **2014**, *3147*, 363–366. [CrossRef]
3. Wang, L.B.; Yan, L.J.; He, A.R. Research and Optimization for Transverse Thickness Difference Control Strategy of UCMW Cold Tandem Mill. *Adv. Mater. Res.* **2013**, *2300*, 1520–1524. [CrossRef]
4. Zhang, Q.D.; Chen, X.L. New Six-High Mills in Cold Tandem Rolling - Control Strategies and Performance Characteristics. *Adv. Mater. Res.* **2012**, *1992*, 1–6. [CrossRef]
5. Du, X.Z.; Yang, Q.; Huang, Q.X. Equation of Shape Control and Edge Drop on UCMW Tandem Cold Mill. *Adv. Mater. Res.* **2012**, *1992*, 94–97. [CrossRef]
6. Zhang, Y.; Yang, Q.; Shao, J. Effect of the end shape of tapered work roll on edge drop of cold rolled strip. In Proceedings of the Third International Conference on Mechanical Engineering and Mechanics, Beijing, China, 21 October 2009; pp. 1411–1416.
7. Laurent, V.D.R.; Jean, J.C.; Stephane, T. New on-line gage for edge drop measurement and effect of tapered work rolls. *Iron Steel Eng.* **1995**, *72*, 27–32.
8. Abdulrahman, A.; Zheng, Y.J.; Dong, B.W. Thin Strip Profile Control Capability of Roll Crossing and Shifting in Cold Rolling Mill. *Mater. Sci. Forum.* **2014**, *2823*, 70–78.
9. Nakhoul, R.; Montmitonnet, P.; Legrand, N. Manifested flatness defect prediction in cold rolling of thin strips. *Int. J. Mater. Form. Off. J. Eur. Sci. Assoc. Mater. Form. ESAFORM* **2015**, *8*, 283–292. [CrossRef]
10. Wang, L.B.; Yan, L.J.; He, A.R. Research on Broken Edge Wave Control Strategy Caused by Edge Drop Control in UCMW Cold Tandem Rolling. *Adv. Mater. Res.* **2013**, *2300*, 1525–1529. [CrossRef]
11. Wang, X.C.; Yang, Q. Research on Asymmetric Shape Control Character and Generalized Overall Shape Control Strategy for Universal Crown Mill. *J. Mech. Eng.* **2012**, *48*, 58–65. [CrossRef]
12. Wehr, M.; Abel, D.; Lohmar, J. Improving the thickness accuracy of cold rolled narrow strip by piezoelectric roll gap control at high rolling speed. *CIRP Ann.* **2018**, *67*, 313–316.
13. Zhang, Q.D.; Wang, W.G.; Wu, Y. Coordinated control of steel strip shape and surface roughness of cold-rolled strip. *Chin. J. Eng.* **2005**, *27*, 232–234.
14. Yan, Q.T.; Zhang, J.; Jia, S.H. Research on Analyzing the Flatness Adjusting Capacity for Cold Mill and Its Application. *J. Mech. Eng.* **2011**, *47*, 77–81. [CrossRef]
15. Han, Z.D.; Liu, X.B.; Yuan, G.Q. Influence Factor Analysis of Loaded Roll Gap Crown with Copper Strip Rolling. *Mach. Des. Manuf.* **2019**, *3*, 234–237, 241.

16. Feng, X.W.; Wang, X.C.; Yang, Q. Analysis of Edge Drop Control Performance of Work Roll Contour on Six-high Cold Rolling Mill. *J. Mech. Eng.* **2019**, *55*, 83–90. [CrossRef]
17. Cao, J.G.; Jiang, J.; Qiu, L. High precision integrated profile and flatness control for new-generation high-tech wide strip cold rolling mills. *J. Cent. South Univ. (Med. Sci.)* **2019**, *50*, 1584–1591.
18. He, A.R.; Shao, J.; Sun, Q. Transverse Thickness Deviation Control of Non-oriented Silicon Steel during Cold Rolling. *J. Mech. Eng.* **2011**, *47*, 25–30. [CrossRef]
19. Cao, J.G.; Zhang, Y.; Yang, G.H. Intermediate roll contour of 6-hi UCM cold rolling mill. *J. Cent. South Univ. (Med. Sci.)* **2011**, *42*, 966–971.
20. Zhou, X.M.; Zhang, Q.D.; Wang, C.S. Analysis of Edge Drop Control Performance and Influencing Factors of UCMW Rolling Mill. *Chin. J. Eng.* **2007**, *29*, 417–420.

metals

Article

Study on Lubrication Characteristics of C4-Alkane and Nanoparticle during Boundary Friction by Molecular Dynamics Simulation

Xuan Zheng [1,*], Lihong Su [2,*], Guanyu Deng [2], Jie Zhang [3], Hongtao Zhu [2] and Anh Kiet Tieu [2]

1 School of Aeronautical Engineering, Taizhou University, Taizhou 318000, China
2 School of Mechanical, Materials, Mechatronic and Biomedical Engineering, University of Wollongong, Wollongong, NSW 2522, Australia; gd577@uowmail.edu.au (G.D.); hongtao@uow.edu.au (H.Z.); ktieu@uow.edu.au (A.K.T.)
3 State Key Laboratory of Tribology, Tsinghua University, Beijing 100084, China; zj8651731@126.com
* Correspondence: zhengxuan@tzc.edu.cn (X.Z.); lihongsu@uow.edu.au (L.S.)

Abstract: Lubricant has been widely applied to reduce wear and friction between the contact surfaces when they are in relative motion. In the current study, a nonequilibrium molecular dynamics (NEMD) simulation was specifically established to conduct a comprehensive investigation on the dynamic contact between two iron surfaces in a boundary friction system considering the mixed C4-alkane and nanoparticles as lubricant. The main research objective was to explore the effects of fluid and nanoparticles addition on the surface contact and friction force. It was found that nanoparticles acted like ball bearings between the contact surfaces, leading to a change of sliding friction mode to rolling friction mode. Under normal loads, plastic deformation occurred at the top surface because nanoparticles were mainly supporting the normal load. By increasing the number of C4-alkane molecules between two contact surfaces, the contact condition has been changed from partial to full lubrication. In addition, an attractive force from the solid–liquid LJ interaction between C4-alkane and surfaces was observed at the early stage of sliding, due to the large space formed by wall surfaces and nanoparticles. The findings in this paper would be beneficial for understanding the frictional behavior of a simple lubricant with or without nanoparticles addition in a small confinement.

Keywords: boundary lubrication; friction contact; molecular dynamics simulation; fluid lubricant; nanoparticles; nanotribology

Citation: Zheng, X.; Su, L.; Deng, G.; Zhang, J.; Zhu, H.; Tieu, A.K. Study on Lubrication Characteristics of C4-Alkane and Nanoparticle during Boundary Friction by Molecular Dynamics Simulation. *Metals* **2021**, *11*, 1464. https://doi.org/10.3390/met11091464

Academic Editor: Slobodan Mitrovic

Received: 1 August 2021
Accepted: 11 September 2021
Published: 15 September 2021

1. Introduction

Recent experiments have found that adding nanoparticles to a lubricant has a significant influence on reducing wear and friction. The types of nanoparticles include metallic [1,2], metallic oxides [3–5] and non-metallic [6–10]. A comprehension discussion and summary on the roles of nanoparticles in oil lubrication has been carried out recently by Dai et al. [11]. They have compared the effects of chemical and physical properties of nanoparticles as well as their morphology and size on the lubricating performances. Due to the chemical composition of nanoparticles, a tribofilm was formed to protect the friction surface [12]. The size of nanoparticles showed visible effects on both friction and wear. The ideal particle size mainly depends on the working condition. Small nanoparticles can easily enter the contact interface, yielding a relatively low friction coefficient [13]. The main morphology of nanoparticles is spherical, followed by sheet, granular, onion and tube, which determines the lubrication conditions which are rolling, sliding, exfoliation, or a combination of all [14]. Moreover, nanoparticles can also affect lubricant flow behavior. A so-called plug flow occurring with nanoparticles and localized shear at surfaces were observed by Sperka and co-authors [15]. In addition, Ghaednia et al. [16] and Jackson et al. [17] have also studied the interactions between nanoparticles with the lubricant or surfaces, and they both found a friction reduction mechanism separately when nanoparticles were present in the lubricants.

However, these mechanisms cannot be directly observed by experiments. Molecular dynamic (MD) simulations provide one of the most cost-effective and precise approaches to investigate molecular or atomistic level phenomenon [18–20]. To date, several simulations based on the molecular dynamics method have been conducted to investigate surface contact and the frictional characteristics of various lubricants and nanoparticles. There are models including single asperity [21], amorphous [22] or 2D rough surface [23], 3D rough surface contact [24], thin film model with flat surface [25] and 3D rough surface [26,27]. It should be pointed out that all these above-mentioned simulations with sliding walls are nonequilibrium molecular dynamics (NEMD) simulations in tribology [28]. According to previous studies [29–35], there is no doubt that NEMD simulations provide unique insights into understanding the nanoscale friction and lubrication mechanisms. For example, Lee et al. [29] studied the rolling resistance of a rigid Ni sphere of Cu substrate. The distribution of contact pressure during indentation was found to be mainly affected by the atomic terraces of the sphere. Their simulations also indicated that smaller nanoparticles and surface with adhesion showed a larger friction coefficient. Joly-Pottuz et al. [30] simulated a rolling-sliding behavior of carbon onions between DLC surfaces, which agreed well with TEM and STM experimental observations. Bucholz et al. [31] found that the interfacial bonds during friction determined the rolling and sliding condition. The NEMD simulations by Eder et al. [32] were about the abrasion process considering a rough iron surface having multiple hard abrasive particles. Ewen et al. [33] developed a model containing two different carbon nanoparticles between iron surfaces and studied the influence of nanoparticle type and coverage. They also discussed the influences of local pressure and velocity on friction. Shi et al. [34] studied the effect of nanoparticle shape on its frictional condition. Their simulation results revealed that the movement pattern turned into rolling from sliding when the nanoparticles were closer to a sphere. Su et al. [35] mixed water and nanodiamonds that had charges as lubricant confined by two gold surfaces. They claimed that the nanodiamonds with negative charges dramatically increased friction force due to the combination of rolling and cutting movement patterns.

However, the reports about the anti-wear mechanisms of fluid containing nanoparticles are still limited. Lv et al. [36] carried out MD simulations of Cu-Ar nanofluids confined within two solid surfaces. It has been found that the nanoparticles exerted a supporting force at the plates and further reduced the contact of two solid surfaces, weakening the friction effect. In addition, with an increase in pressure, nanoparticles were cut and absorbed into the solid surface that possessed a potential influence to fill for rough plates. Ji et al. [37] performed MD simulations to investigate the roles of nanoparticles on the sliding friction process. According to their simulation results, nanoparticles had a filling effect to smooth the contact surface and further to reduce the friction force. Hu et al. [38,39] studied the effect of nanoparticles by considering interactions with the base fluid. They observed a higher transition pressure due to the presence of nanoparticles. Moreover, they also compared the effect of the diamond and SiO_2 nanoparticles during sliding. Specifically, hard nanoparticles were found to polish the friction surfaces.

In the boundary and mixed lubrication conditions, direct metal/metal contacts often occur due to less liquid lubricants [40]. Therefore, it is very critical to explore the effect of fluid combining with nanoparticles on preventing direct surfaces contact. In our previous reports [25–27], we have systematically studied the influence of surface roughness during dry or lubricated contact conditions. The current work is an extension of these studies. NEMD simulations considering the lubricating conditions between two iron wall surfaces using the mixed C4-alkane and nanoparticles is specifically performed to investigate the influences of fluid and nanoparticle and associated mechanisms on surfaces contact and friction force evolution. This study provides a fundamental understanding on the roles of mixed fluid and nanoparticles in the boundary lubrication system that will be beneficial for understanding the frictional behavior of a simple lubricant with or without adding nanoparticles in a small confinement.

2. Materials and Methods

2.1. Materials and Model Setup

The MD simulation is performed through a large scale Atomic/Molecular Massively Parallel Simulator (LAMMPS) [41], which has been proven to be a very powerful MD simulator. As displayed in Figure 1, the geometry model consists of two <001> oriented BCC iron upper and lower walls, randomly distributed C4 molecules and 16 iron rigid nanoparticles with radius 8 Å, which leads to about 19% surface coverage by the nanoparticles. It should be noted that the use of iron for walls and nanoparticles is a simplification in this study. The iron wall is divided further into three regions: rigid region, thermostat region and free deformable region. Periodic boundary conditions have been defined along the X/Y lateral directions. In contrast, the Z normal direction is variable when the system height fluctuates; the system is compressed and sheared under 4 normal loads of 0.25, 0.5, 0.75 and 1.0 GPa. The lower and upper iron walls slide against each other with a velocity of 20 m/s in the X direction, which results in the lubricant shearing at 40 m/s. It should be stated that the relative motion velocity of 40 m/s is quite high compared to the practical applications and real experiments; however, it is mainly used in this study to model several sliding cycles within a reasonable simulation time. In the current investigation, different amounts of C4-alkane molecular are used to construct lubricant flow, which could provide a different lubricated condition, such as less or full lubrication.

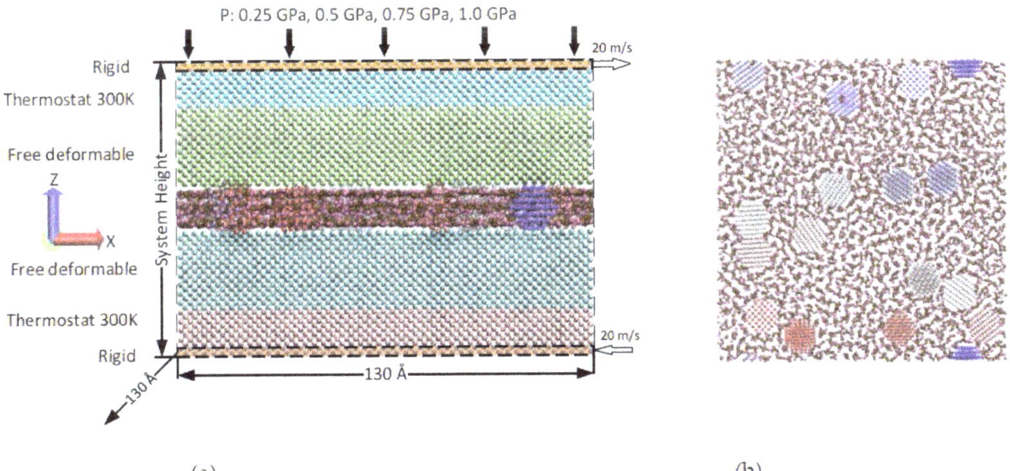

Figure 1. (**a**) A typical molecular structure showing C4-alkane and 16 nanoparticles were confined between the iron walls and (**b**) top view of nanoparticles and C4-alkane fluid. Normal load (P) is only applied on the upper rigid region, but the lower rigid region is fixed in the Z direction. These white arrows exhibit the sliding direction at a speed of 20 ms/s imposed on rigid regions.

2.2. Force Field

In this work, there might be a severe plastic deformation that intensively depends on the proper force field, which could not be handled correctly through simple potentials including the Morse and Lennard–Jones potentials. Therefore, the embedded-atom method (EAM) force field designed for solid Fe was used in this work [42].

$$U(r_{ij}) = 4\varepsilon\left[\left(\frac{\sigma}{r_{ij}}\right)^{12} - \left(\frac{\sigma}{r_{ij}}\right)^{6}\right] \quad (1)$$

$$\sigma_{\mathrm{ij}} = \frac{1}{2}(\sigma_{\mathrm{ii}} + \sigma_{\mathrm{jj}}) \;\; \varepsilon_{\mathrm{ij}} = \sqrt{\varepsilon_{\mathrm{ii}}\varepsilon_{\mathrm{jj}}} \quad (2)$$

$$E_{\mathrm{bond}} = \frac{K_b}{2}(r - r_0)^2 \quad (3)$$

$$E_{\mathrm{bend}} = \frac{K_\theta}{2}(\theta - \theta_0)^2 \quad (4)$$

$$E_{\mathrm{torsion}} = c_0 + c_1[1 + \cos(\alpha)] + c_2[1 - \cos(2\alpha)] + c_3[1 + \cos(3\alpha)] \quad (5)$$

The C4-alkane is simulated by the TraPPE United Atom force field which utilizes pseudo-atoms to represent the CHx groups [43]. It should be noted that, TraPPE United Atom for field is not the best one to accurately describe the material properties compared to other advanced force fields, but it is still widely used from the consideration of computational efficiency [25–27,43–45]. Thus, we have applied the TraPPE United Atom force filed for atomistic simulations in order to reduce the computational cost, which is very important for NEMD simulations when there are a huge number of cases to run. The bond stretching of E_{bond}, angle bending of E_{bend} and torsion angle of E_{torsion} have been calculated by Equations (3)–(5), respectively. The Lennard–Jones potential is used to model the interactions between the segment-pairs that are separated by more than three bonds as displayed in Equation (1). The interactions between surface–particle, liquid–particle, and particle–particle are also calculated by LJ potential. The parameters of unlike interactions were calculated by the Lorentz–Berthelot combining laws; see Equation (2). All the TraPPE-UA and LJ potential parameters are listed in Table 1.

Table 1. The corresponding potential parameters used for alkane and Fe in this study [43,46,47].

LJ 12-6 Potentials	**σ (Å)**		**ε(eV)**	**Mass (g/mol)**
CH3	3.75		0.008444	15.0351
CH2	3.95		0.003963	14.0272
CH	4.68		0.000861	13.0191
Fe * [46]	2.321		0.02045	55.8450
Bond	K_b (eV/ Å2)		r_0 (Å)	
C-C [47]	39.0279464		1.54	
Angle	K_θ (eV/rad2)		θ_0 (degrees)	
C-C-C	5.3858393		114	
Dihedral	c_0 (eV)	c_1 (eV)	c_2(eV)	c_3 (eV)
C-C-C-C	0	0.030594	−0.005876	0.068190

* —LJ parameters of Fe are for surface–particle, liquid–particle and particle–particle interactions.

3. MD Simulation Procedures

The detailed model geometry has been introduced in the previous section. Each MD simulation is divided into three steps. The first step is dynamic relaxation with 200 ps. The atoms of both lower and upper rigid layers are constrained to maintain the system height, which allows these lubricant molecules to fully relax and the model to reach equilibrium with no variation in the energy of the system. In the second step, uniformly distributed loads of 0.25, 0.5, 0.75 and 1.0 GPa are added on the atoms belonging to the upper rigid layer. On the other hand, the atoms of the lower rigid layer are fixed along the normal direction, and therefore the lubricant is compressed. However, the rigid layers are limited within a maximum distance of 0.00025 Å every time step during the limited compression to achieve gentile touch between deformable layers and rigid particles. Within this stage, in order to maintain the temperature at 300 K, the Nose–Hoover thermostat [48] is applied on the atoms of thermostat layers with a damping coefficient of 100 fs. After the limited compression for 600 ps, the rigid layers are updated freely at each time step for the subsequent 200 ps. The last step is a sliding or shearing process. The atoms of upper rigid layer are sliding at a velocity of 20 m/s along the X direction for 3000 ps. The same sliding velocity is applied to the atoms of lower rigid layer but in an opposite direction, which results in the lower and upper walls sliding and lubricant shearing. During this step, the energy is dissipated via thermostat regions. In order to conduct the numerical integration of atomic classical equations of motions, the Velocity–Verlet algorithm is used in the present study, and the time step is set to 2 fs. The total simulation time was 4000 ps. For the post-processing of simulation data, both friction force and system height are saved every 4 ps, and a smooth window with size of 20 is used for figure plotting.

4. Results and Discussion

During the dynamic relaxation step, different numbers of C4-alkane from 500 to 4000 molecules are used as lubricants. Figure 2 exhibits a typical example that 1500 C4-alkane molecules and 16 rigid particles are fully relaxed during the first 200 ps. All the nanoparticles are close to the lower wall. This is because their initial positions are closer to the lower wall than those of the upper wall. For simplicity, flat model, particle model and lubricated model are used in the subsequent sections to present the model without nanoparticles and C4-alkane molecules, the model only lubricated with nanoparticles, and the model lubricated with a mixture of nanoparticles and C4-alkane molecules, respectively.

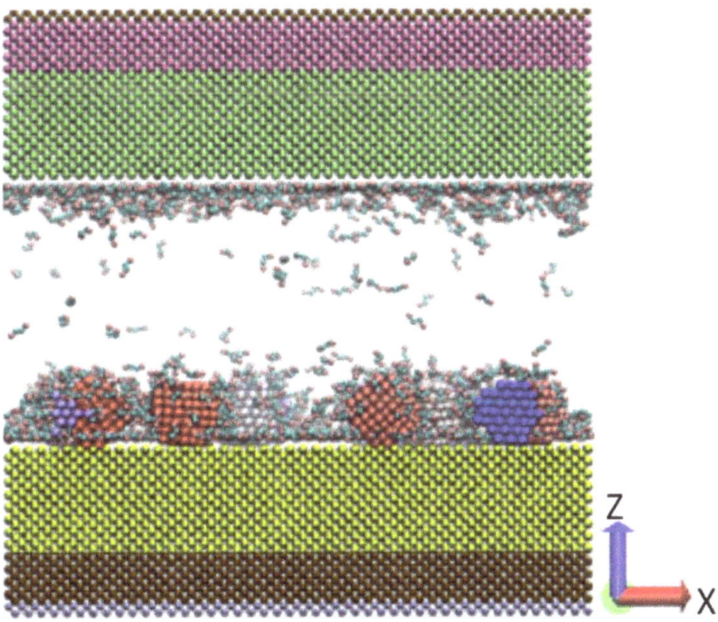

Figure 2. Lubricated model with 1500 C4-alkane molecules in the relaxation step.

4.1. Dry Contact

Aiming to study the influence of nanoparticles during boundary friction, studies on the flat and particle models are conducted first. After relaxation, different normal loads varying from 0.25 to 1.0 GPa are applied on iron atoms of the upper rigid layer. Figure 3 indicates the system height evolution against the simulation time, and a comparison between the flat and nanoparticle models is performed. It is evident from the figure that there is no obvious system height change for the flat model under all normal loads. This is due to the high stiffness of pure Fe. During the subsequent sliding process, a slight fluctuation of system height is observed, which is due to the same surface structures of both upper and lower walls. In contrast, much more stable system heights are observed in the nanoparticle models when the normal load varies between 0.25 and 0.75 GPa. However, when the normal load is increased up to 1.0 GPa, there is an obvious drop in the system height. Moreover, the drop becomes more severe with the sliding distance, which indicates that plastic deformation may occur at the top surfaces of both upper and lower walls. More details will be discussed in the later section of this paper.

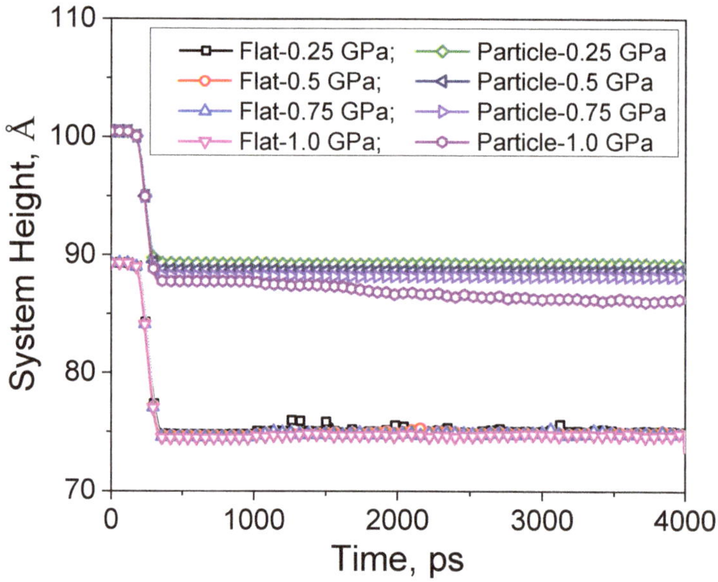

Figure 3. Comparison of the system height evolution history against the simulation time under various normal loads between the flat and particle models.

Although the system height predicted by the flat models in Figure 3 remains stable, the corresponding simulated friction force is relatively large and fluctuates dramatically during the sliding process. Figure 4a,b reveals the influence of normal load on the evolution history of friction force against the simulation time obtained in different simulation systems. For flat models (Figure 4a), the upper and lower surfaces contact directly. There is an unusual friction force change when the simulation time is increased from 250 to 500 ps, as marked by the red dashed circle in the figure. This phenomenon is mainly due to the alignment pattern between the upper and lower walls. Similar jump in the friction force was also observed by Stephan et al. [49], in which the simulated tangential force increased as the chip started to form at the beginning of the lateral movement. On the other hand, An and co-authors [50] have successfully proposed a quantitative relationship of the nanoscale friction force and coefficient between liquid and solid based on their atomic force microscopy experimental characterizations. To better illustrate this effect, a typical system slice along the vertical direction is shown in Figure 5. At 300 ps, the upper wall normally approaches the lower wall. Due to the repulsive force between two deformable layers, the force Fx is increased. However, the force decreases immediately when two surfaces have the shortest distance because the walls can move freely along both X and Y directions.

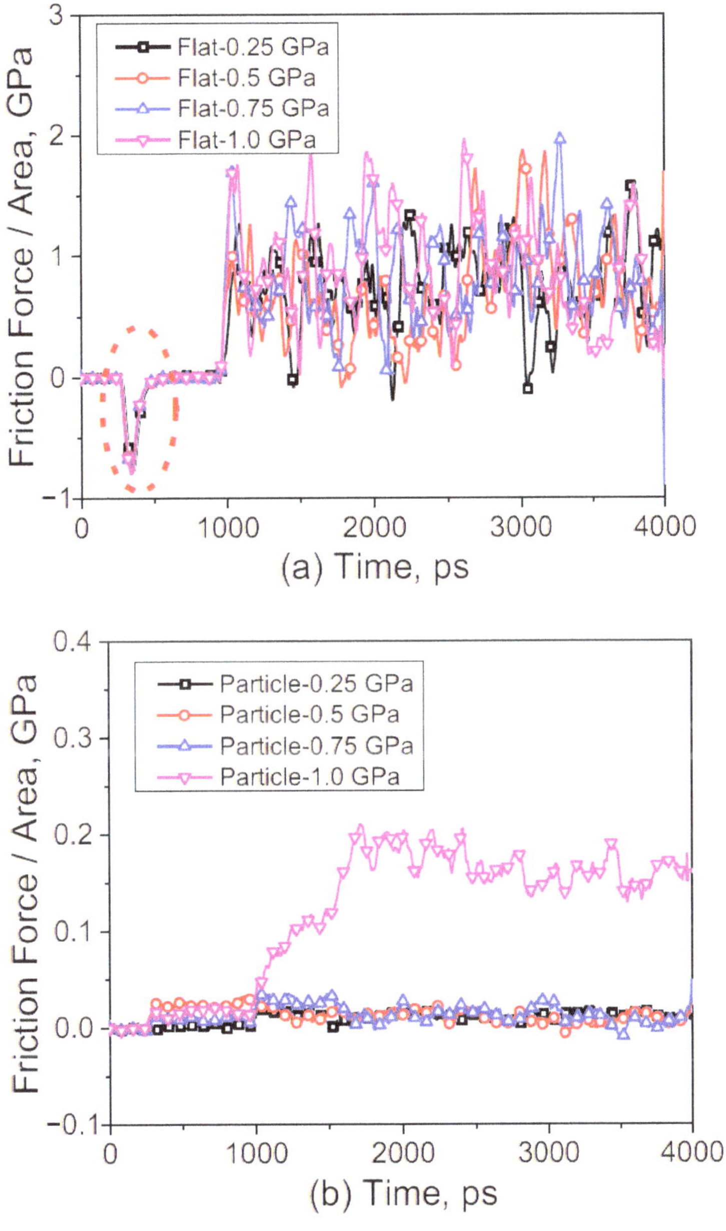

Figure 4. The friction force evolution history when normal load is increased from 0.25 to 1.0 GPa in the (**a**) flat and (**b**) particle models.

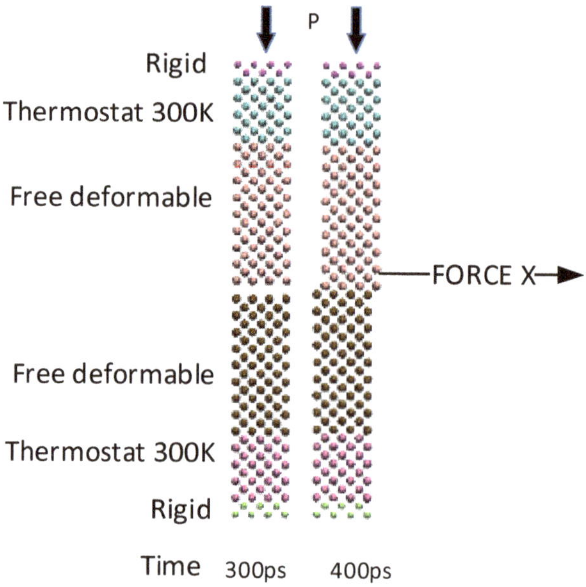

Figure 5. The typical system slices (70 Å < x < 80 Å) at the moment of 300 and 400 ps. The unusual force Fx is due to the change of the upper and lower walls from initial mismatch pattern to match pattern.

In comparison with the results shown in Figure 4a by flat models, the particle models predict much smaller friction forces as displayed in Figure 4b for all the studied normal loads. The wall blocks separating from each other are observed when the hard nanoparticles are added in the boundary friction system. This agrees well with the observations by Tao et al. [51] that nanoparticles are able to divide the rubbing planes and thereby avoid direct surface contact. Specifically, while the normal load is $\leq$0.75 GPa, the friction force drops nearly to zero. This is likely owing to the rolling effect of nanoparticles when the normal load is very low [39]. Figure 6 reveals that the rolling effect of nanoparticles helps to decrease the friction force. For the normal load of 0.25 GPa, the nanoparticles are supporting the upper surface against the normal load but also smoothly rolling between the surfaces during the sliding process. The shape of atoms in gray color remains unchanged through the whole sliding process, suggesting the absence of plastic deformation in the friction pairs. When the normal load rises to 1.0 GPa, there is an obvious rise in the friction force as can be seen from Figure 4b. It means that the nanoparticles induce the plastic deformation in the top surface and thus lead to the increase of friction force. This deformation is shown in Figure 7. It should be noted that no nanoparticles are shown in the figure for clearness. The atoms of the top surface are deformed to accommodate the rigid nanoparticles, which is to ensure the rolling effect between surfaces.

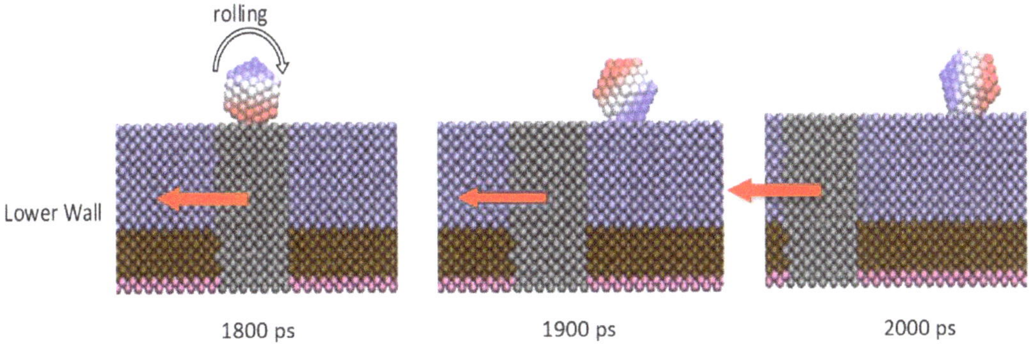

Figure 6. Snapshots of particle models under 0.25 GPa at the moment of 1800, 1900 and 2000 ps. Only the lower wall and one typical particle are shown. Atoms in the particle are colored by their Z position to demonstrate the rolling effect. Atoms in Gray are in the slice from 100 Å to 120 Å along the X direction. The red arrow indicates the slice movement.

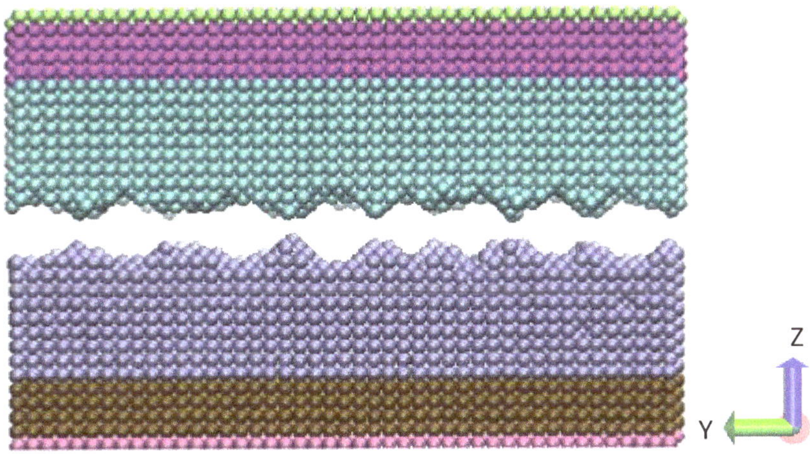

Figure 7. A side view of the particle model with 16 nanoparticles under the normal load of 1.0 GPa during sliding at 4000 ps. Only the upper and lower walls are shown to illustrate the plastic deformation during the sliding process.

In order to gain a better understanding on the plastic deformation under boundary friction condition, particle models with different numbers of nanoparticles are also carried out in this study. Specifically, four cases with 4, 8, 12 and 16 nanoparticles placed between the friction surfaces are considered. Moreover, only the normal load of 1.0 GPa is applied in these cases to ensure that the nanoparticles are able to penetrate the wall blocks. Compared with the flat model, the addition of 4 nanoparticles results in a slight increase of system height, but it remains stable during sliding as can be seen from Figure 8. Although additions of 12 and 16 nanoparticles lead to almost the same system height during compression step, the system height of the model containing 12 nanoparticles drops more quickly once the sliding starts. It has been found that fewer nanoparticles share larger compression stress and thus penetrate deeper into the wall blocks. This phenomenon becomes more obvious when the model contains 8 nanoparticles. In order to understand the influence of the number of nanoparticles, the friction force obtained in the model containing 4, 8, 12 or 16 nanoparticles is plotted in Figure 9, respectively. A comparison clearly indicates that more nanoparticles ensure the rolling effect and further reduce the friction force.

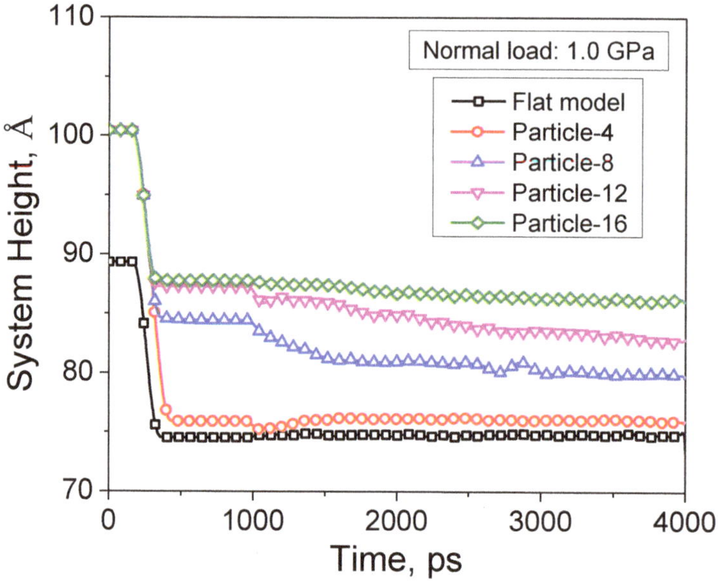

Figure 8. Comparison of the evolution of system heights against the simulation time between the flat model and particle models containing 4, 8, 12 and 16 nanoparticles.

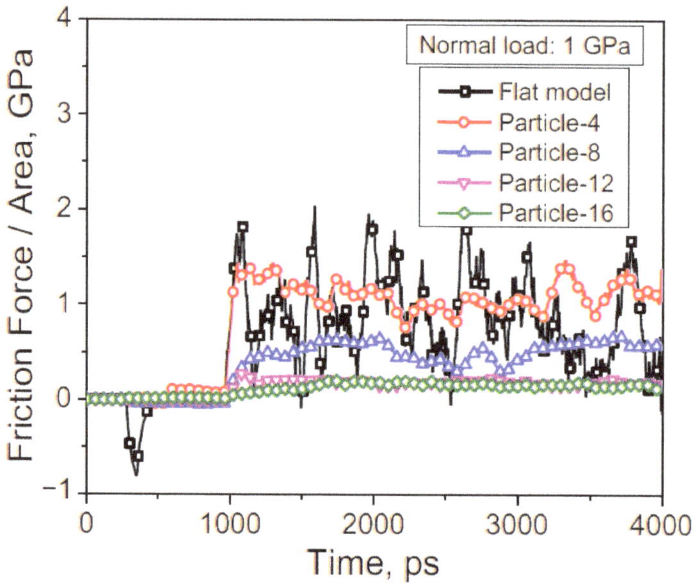

Figure 9. Comparison of the evolution of friction force against the simulation time between the flat model and particle models containing 4, 8, 12 and 16 nanoparticles.

Figure 10 shows the surface morphology of the lower wall after sliding process. It is evident that grooves form during friction, which indicates that there is a cutting action from nanoparticles on surfaces. Such kind of grooves or scratches formation on the worn surface has often been experimentally observed in either nanoscale or microscale [52,53]. It has to be mentioned that only atoms of the lower wall with $Z > 25$ Å are shown in

Figure 10 to clearly demonstrate the cutting effect. The corresponding color bar indicates the Z position of the atoms of the lower wall. As displayed in Figure 10a, all 4 nanoparticles deeply penetrate into the wall blacks under the normal load of 1.0 GPa. In addition, the atoms next to nanoparticles tend to accumulate and act like a cutting tool to remove the materials during the sliding process. Very similar results are also found in the model with 8 nanoparticles, as displayed in Figure 10b. By increasing the number of nanoparticles to 12 and 16, more grooves are formed with smaller height difference at block surfaces, as can be seen from Figure 10c,d, respectively.

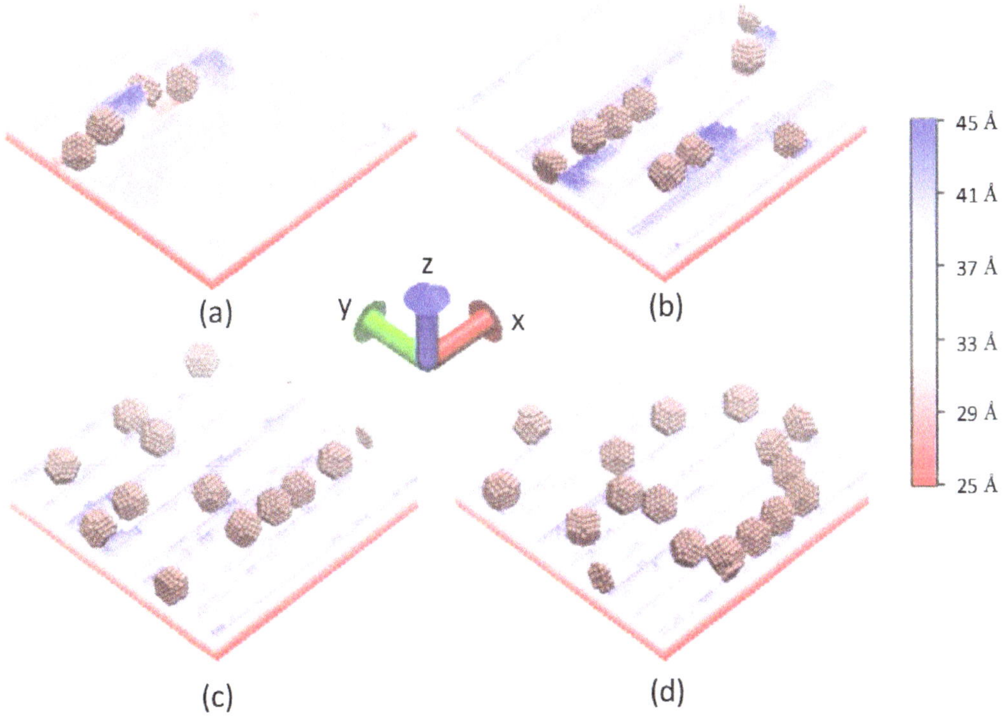

Figure 10. The surface morphology of the lower wall with (**a**) 4, (**b**) 8, (**c**) 12 and (**d**) 16 nanoparticles at 4000 ps. Only the atoms of the lower wall with Z > 25 Å are shown and colored. The normal load applied in the boundary friction system is 1.0 GPa.

4.2. Lubricated Contact

When lubricant is present in the boundary friction system, the wall surfaces are subjected to a combined influence of nanoparticles, adhesion as well as the lubricant, particularly while the average thickness of the lubricant is comparable to the surface roughness and the size of nanoparticles [23]. In this section, MD simulations considering the C4-alkane as the lubricant are carried out to investigate its influence on the boundary friction.

Figure 11 displays the system height evolution for the lubricated models with C4-alkane and 16 nanoparticles. Each system has an initial system height, which is apparently dependent on the number of C4-alkane molecules. However, all systems have the same height reduction rate because of the limited distance movement at each time step. This step could ensure the gentile touch between two wall surfaces and nanoparticles. Before the sliding process, it is important to guarantee the initial contact surface. Otherwise, the sharp contact could result in severe indentation on the wall surface, remaining in the following sliding process. From Figure 11, it has been found that C4-alkane molecules less than 1000 are unable to support the upper wall as the height does not change compared with the particle model. However, 2000 molecules of C4-alkane start supporting against the normal load as the height increases. More obvious increases could be observed in the systems which have 3000 or 4000 molecules.

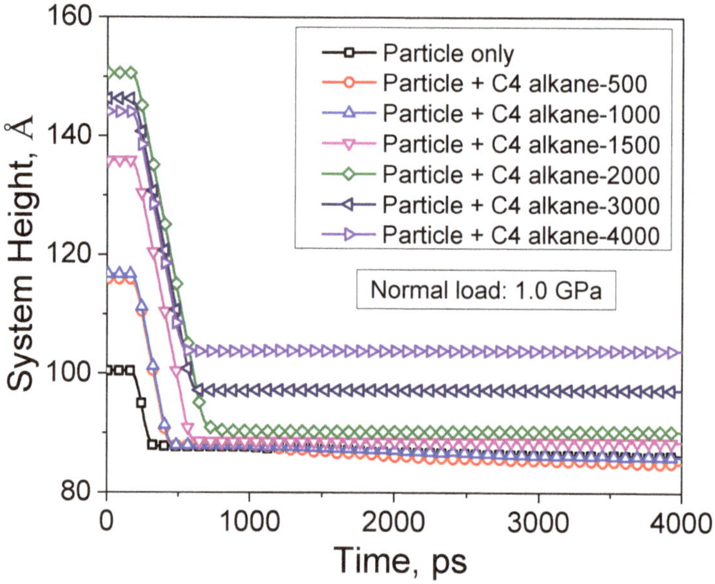

Figure 11. Comparison of the system height evolution history against the simulation time between the particle and lubricated models under the normal load of 1.0 GPa. The number of C4-alkane molecules varies from 500 to 4000.

Similar to the system height evolution, the friction force also varies when the number of lubricant molecules changes. From the results shown in Figure 12, it can be found that C4-alkane with 500 and 1000 molecules has nearly no contributions as a lubricant to decrease the friction force. However, when the number of molecules is greater than 1500 or above, an obvious change in the friction force is visible. It is due to the flow of lubricant starting to support the upper wall against the normal load. Therefore, the particles are subjected to less compression. It seems that the surfaces are only lubricated by fluid, and similar results were also observed in our previous study [25]. It is worth noting that, in addition to the system height, the nanoparticle-surface indentation depth is also a very important parameter which makes it possible to link and compare the simulation and experimental results [54,55]. Such kind of analysis will be conducted in our future work, where a series of new NEMD simulations considering different types of nanoparticles with 3D rough surfaces will be developed.

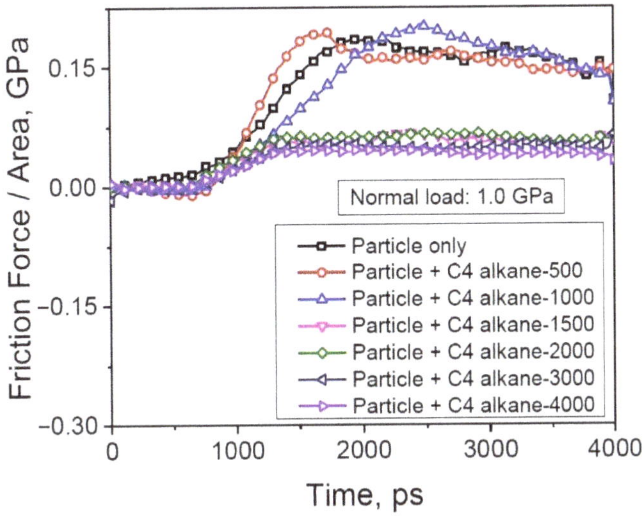

Figure 12. Comparison of the friction force evolution history against the simulation time between the particle and lubricated models under the normal load of 1.0 GPa. The number of C4-alkane molecules varies from 500 to 4000.

To better explore the effect and mechanism of the lubricant fluid in the boundary lubrication system, the component of friction force corresponding to the systems having 500, 1000, 1500 and 2000 molecules is plotted in Figure 13, respectively. When the system only has 500 C4-alkane molecules mixed with nanoparticles, the supporting force is mainly contributed by the nanoparticles, as can be seen from Figure 13a. There is no direct supporting force from the lower wall as the gap distance is larger than the LJ cutoff distance. It is interesting that the force Z from C4-alkane decreases with the sliding process and even becomes negative after 1500 ps. It means that C4-alkane helps to compress the upper and lower walls. This is likely due to the large space formed by wall surfaces and nanoparticles. In addition, C4-alkane is free to flow and naturally the molecules tend to accumulate. The adsorption between C4-alkane and surfaces finally results in an attractive force between them. However, the negative force from C4-alkane disappears quickly when the flow is under compression or there is less space, which could be observed in Figure 13b. During the compression step and early stage of sliding, the supporting force from C4-alkane is near zero. It starts to increase at 2500 ps as the system height decreases slightly from then on. When 1500 molecules are mixed with nanoparticles, the force components remain stable during the sliding process. However, on the other hand, 2000 molecules exhibit a completely different influence on the supporting force. Figure 13d shows that nanoparticles exert an attractive force on wall surfaces, which results in a larger compression force on the lubricant flow. In this condition, nanoparticles could move freely within the lubricant flow. However, the nanoparticle atoms are still in the cut-off distance with surface atoms, which finally results in an attractive force. Therefore, these results reveal a great influence of the number of C4-alkane molecules on its lubricity within a boundary friction system and a transition from partial to full lubrication.

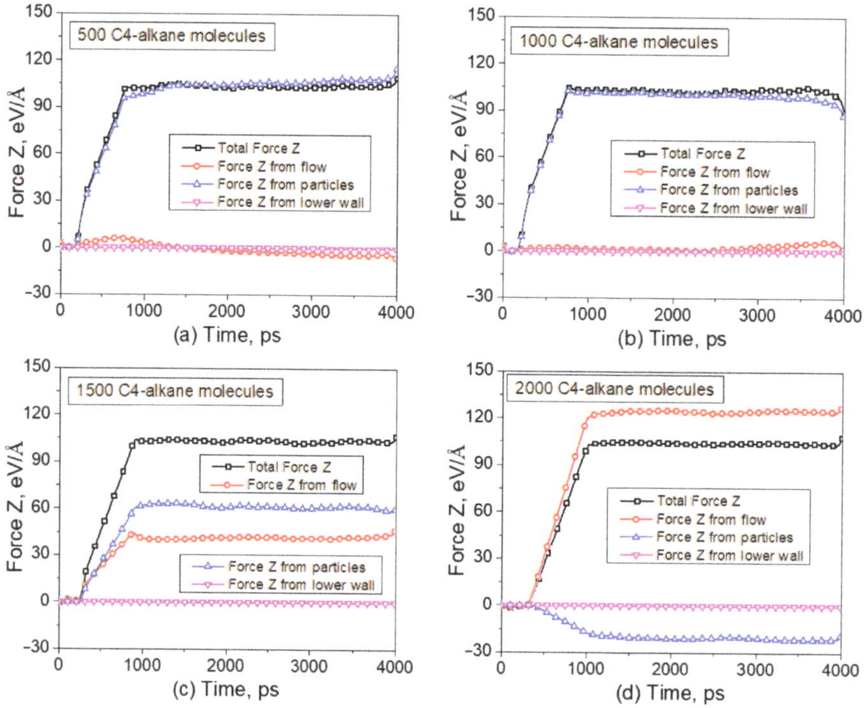

Figure 13. Influence of the number of C4-alkane molecules on the total force Z and its component from flow, nanoparticles and the lower wall: (**a**) 500, (**b**) 1000, (**c**) 1500 and (**d**) 2000 molecules.

5. Conclusions

In the present study, a comprehensive NEMD simulation model considering two iron wall surfaces, mixed C4-alkane and nanoparticles as lubricant has been established to explore the influences of nanoparticles and fluid on the surface contact and friction force during boundary friction system. The simulation results revealed that nanoparticles were acting like ball bearings between contact surfaces, leading to a change from the sliding friction to rolling friction. When a large normal load greater than 0.75 GPa was applied, there was severe plastic deformation at the top surface. Moreover, C4-alkane with a number of molecules lower than 1500 was unable to support the upper wall, which had a similar friction behavior to that of the cases with only nanoparticles. In addition, there was an attractive force from C4-alkane molecules to surfaces, which resulted in an extra compression. This was due to the large space formed by wall surfaces and nanoparticles during the sliding process. However, with the increase of the number of molecules, the surfaces were gradually separated by the lubricant fluid from the mixed lubrication to totally thin film lubrication, and the friction force dropped dramatically.

Author Contributions: Conceptualization, X.Z. and L.S.; methodology, X.Z., L.S. and G.D.; software, X.Z., G.D. and J.Z.; validation, J.Z. and H.Z.; investigation, X.Z., L.S. and G.D.; resources, L.S. and A.K.T.; writing—original draft preparation, X.Z., L.S., G.D. and J.Z.; writing—review and editing, X.Z., L.S., H.Z. and A.K.T.; project administration, L.S.; funding acquisition, L.S. and A.K.T. All authors have read and agreed to the published version of the manuscript.

Funding: This research was funded by Australian Research Council (Grant Numbers: DE180100124, DP170103173 and DP190103455).

Data Availability Statement: Not applicable.

Acknowledgments: All authors appreciate very much for conducing simulations using a HPC cluster that belonged to the University of Wollongong and the computing facilities provided by NCI National Facility of Australia. L.S. is grateful to the Australian Research Council for awarding her the Discovery Early Career Researcher Award (DECRA) fellowship (DE180100124). G.D., H.Z. and K.T. would like to acknowledge the financial supports from the Australian Research Council (DP 170103173 and DP190103455).

Conflicts of Interest: The authors declare no conflict of interest.

References

1. Zhou, J.; Wu, Z.; Zhang, Z.; Liu, W.; Xue, Q. Tribological behaviour and lubricating mechanism of Cu nanoparticles in oil. *Tribol. Lett.* **2000**, *8*, 213–218. [CrossRef]
2. Qiu, S.; Zhou, Z.; Dong, J.; Chen, G. Preparation of Ni nanoparticles and evaluation of their tribological performance as potential antiwear additives in oils. *J. Tribol.* **2001**, *123*, 441–443. [CrossRef]
3. Xue, Q.; Liu, W.; Zhang, Z. Friction and wear properties of a surface-modified TiO_2 nanoparticle as an additive in liquid paraffin. *Wear* **1997**, *213*, 29–32. [CrossRef]
4. Hu, Z.S.; Dong, J.X.; Chen, G.X. Study on antiwear and reducing friction additive of nanometer ferric oxide. *Tribol. Int.* **1998**, *31*, 355–360. [CrossRef]
5. Radice, S.; Mischler, S. Effect of electrochemical and mechanical parameters on the lubrication behaviour of Al_2O_3 nanoparticles in aqueous suspensions. *Wear* **2006**, *261*, 1032–1041. [CrossRef]
6. Gupta, B.K.; Bhushan, B. Fullerene particles as an additive to liquid lubricants and greases for low friction and wear. *Lubr. Eng.* **1994**, *50*, 524–528.
7. Cizaire, L.; Vacher, B.; Le Mogne, T.; Martin, J.M.; Rapoport, L.; Margolin, A.; Tenne, R. Mechanisms of ultra-low friction by hollow inorganic fullerene-like MoS2 nanoparticles. *Surf. Coat. Technol.* **2002**, *160*, 282–287. [CrossRef]
8. Rabaso, P.; Ville, F.; Dassenoy, F.; Diaby, M.; Afanasiev, P.; Cavoret, J.; Vacher, B.; Le Mogne, T. Boundary lubrication: Influence of the size and structure of inorganic fullerene-like MoS_2 nanoparticles on friction and wear reduction. *Wear* **2014**, *320*, 161–178. [CrossRef]
9. Tenne, R. Inorganic nanotubes and fullerene-like nanoparticles. *Nat. Nanotechnol.* **2006**, *1*, 103–111. [CrossRef] [PubMed]
10. Lahouij, I.; Dassenoy, F.; Vacher, B.; Martin, J.M. Real time TEM imaging of compression and shear of single fullerene-like MoS_2 nanoparticle. *Tribol. Lett.* **2011**, *45*, 131–141. [CrossRef]
11. Dai, W.; Kheireddin, B.; Gao, H.; Liang, H. Roles of nanoparticles in oil lubrication. *Tribol. Int.* **2016**, *102*, 88–98. [CrossRef]
12. Xie, H.; Jiang, B.; He, J.; Xia, X.; Pan, F. Lubrication performance of MoS_2 and SiO_2 nanoparticles as lubricant additives in magnesium alloy-steel contacts. *Tribol. Int.* **2016**, *93*, 63–70. [CrossRef]
13. Moshkovith, A.; Perfiliev, V.; Lapsker, I.; Fleischer, N.; Tenne, R.; Rapoport, L. Friction of fullerene-like WS_2 nanoparticles: Effect of agglomeration. *Tribol. Lett.* **2006**, *24*, 225–228. [CrossRef]
14. Tevet, O.; Von-Huth, P.; Popovitz-Biro, R.; Rosentsveig, R.; Wagner, H.D.; Tenne, R. Friction mechanism of individual multilayered nanoparticles. *Proc. Natl. Acad. Sci. USA* **2011**, *108*, 19901–19906. [CrossRef]
15. Šperka, P.; Křupka, I.; Hartl, M. Evidence of plug flow in rolling—sliding elastohydrodynamic contact. *Tribol. Lett.* **2014**, *54*, 151–160. [CrossRef]
16. Ghaednia, H.; Babaei, H.; Jackson, R.L.; Bozack, M.J.; Khodadadi, J.M. The effect of nanoparticles on thin film elasto-hydrodynamic lubrication. *Appl. Phys. Lett.* **2013**, *103*, 263111. [CrossRef]
17. Jackson, R.L.; Ghaednia, H.; Babaei, H.; Khodadadi, J.M. Comment on Sperka, P., I. Krupka, M. Hartl (2014). "evidence of plug flow in rolling—sliding elastohydrodynamic contact." Tribology Letters 54(2): 151–160. *Tribol. Lett.* **2014**, *56*, 407. [CrossRef]
18. Srivastava, I.; Kotia, A.; Ghosh, S.K.; Ali, M.K.A. Recent advances of molecular dynamics simulations in nanotribology. *J. Mol. Liq.* **2021**, *335*, 116154. [CrossRef]
19. Zhang, J.; Su, L.; Wang, Z. Concurrent multiscale simulations of rough lubricated contact of aluminum single crystal. *Metals* **2020**, *10*, 965. [CrossRef]
20. Stephan, S.; Lautenschlaeger, M.P.; Alhafez, I.A.; Horsch, M.T.; Urbassek, H.M.; Hasse, H. Molecular dynamics simulation study of mechanical effects of lubrication on a nanoscale contact process. *Tribol. Lett.* **2018**, *66*, 126. [CrossRef]
21. Eder, S.; Vernes, A.; Vorlaufer, G.; Betz, G. Molecular dynamics simulations of mixed lubrication with smooth particle post-processing. *J. Phys. Condens. Matter* **2011**, *23*, 175004. [CrossRef]
22. Jabbarzadeh, A.; Harrowell, P.; Tanner, R.I. Low friction lubrication between amorphous walls: Unraveling the contributions of surface roughness and in-plane disorder. *J. Chem. Phys.* **2006**, *125*, 034703. [CrossRef]
23. Jabbarzadeh, A.; Atkinson, J.D.; Tanner, R.I. Effect of the wall roughness on slip and rheological properties of hexadecane in molecular dynamics simulation of Couette shear flow between two sinusoidal walls. *Phys. Rev. E* **2000**, *61*, 690–699. [CrossRef]
24. Spijker, P.; Anciaux, G.; Molinari, J.F. Dry sliding contact between rough surfaces at the atomistic scale. *Tribol. Lett.* **2011**, *44*, 279–285. [CrossRef]
25. Zheng, X.; Zhu, H.; Tieu, A.K.; Chen, K. Molecular dynamics simulation of confined n-alkanes: Ordered structure and crystalline bridges. *Int. J. Surf. Sci. Eng.* **2014**, *8*, 201–212. [CrossRef]

26. Zheng, X.; Zhu, H.; Tieu, A.K.; Kosasih, B. A molecular dynamics simulation of 3D rough lubricated contact. *Tribol. Int.* **2013**, *67*, 217–221. [CrossRef]
27. Zheng, X.; Zhu, H.; Tieu, A.K.; Kosasih, B. Roughness and lubricant effect on 3D atomic asperity contact. *Tribol. Lett.* **2013**, *53*, 215–223. [CrossRef]
28. Ewen, J.P.; Heyes, D.M.; Dini, D. Advances in nonequilibrium molecular dynamics simulations of lubricants and additives. *Friction* **2018**, *6*, 349–386. [CrossRef]
29. Lee, W.G.; Cho, K.H.; Jang, H. Molecular dynamics simulation of rolling friction using nanosize spheres. *Tribol. Lett.* **2009**, *33*, 37–43. [CrossRef]
30. Joly-Pottuz, L.; Bucholz, E.W.; Matsumoto, N.; Phillpot, S.R.; Sinnott, S.B.; Ohmae, N.; Martin, J.M. Friction properties of carbon nano-onions from experiment and computer simulations. *Tribol. Lett.* **2010**, *37*, 75–81. [CrossRef]
31. Bucholz, E.W.; Phillpot, S.R.; Sinnott, S.B. Molecular dynamics investigation of the lubrication mechanism of carbon nano-onions. *Comput. Mater. Sci.* **2012**, *54*, 91–96. [CrossRef]
32. Eder, S.J.; Feldbauer, G.; Bianchi, D.; Cihak-Bayr, U.; Betz, G.; Vernes, A. Applicability of macroscopic wear and friction laws on the atomic length scale. *Phys. Rev. Lett.* **2015**, *115*, 025502. [CrossRef]
33. Ewen, J.P.; Gattinoni, C.; Thakkar, F.M.; Morgan, N.; Spikes, H.A.; Dini, D. Nonequilibrium molecular dynamics investigation of the reduction in friction and wear by carbon nanoparticles between iron surfaces. *Tribol. Lett.* **2016**, *63*, 38. [CrossRef]
34. Shi, J.; Fang, L.; Sun, K. Friction and wear reduction via tuning nanoparticle shape under low humidity conditions: A nonequilibrium molecular dynamics simulation. *Comput. Mater. Sci.* **2018**, *154*, 499–507. [CrossRef]
35. Su, L.; Krim, J.; Brenner, D.W. Dynamics of Neutral and charged nanodiamonds in aqueous media confined between gold surfaces under normal and shear loading. *ACS Omega* **2020**, *5*, 10349–10358. [CrossRef] [PubMed]
36. Lv, J.; Bai, M.; Cui, W.; Li, X. The molecular dynamic simulation on impact and friction characters of nanofluids with many nanoparticles system. *Nanoscale Res. Lett.* **2011**, *6*, 200. [CrossRef] [PubMed]
37. Ji, C.; Sun, S.; Wang, B.; Lin, B. Molecular dynamic simulations of the roles of nanoparticles in sliding friction process. *Chem. Phys. Lett.* **2019**, *728*, 44–49. [CrossRef]
38. Hu, C.; Bai, M.; Lv, J.; Wang, P.; Li, X. Molecular dynamics simulation on the friction properties of nanofluids confined by idealized surfaces. *Tribol. Int.* **2014**, *78*, 152–159. [CrossRef]
39. Hu, C.Z.; Bai, M.L.; Lv, J.Z.; Kou, Z.H.; Li, X.J. Molecular dynamics simulation on the tribology properties of two hard nanoparticles (diamond and silicon dioxide) confined by two iron blocks. *Tribol. Int.* **2015**, *90*, 297–305. [CrossRef]
40. Persson, B.N.J.; Samoilov, V.N.; Zilberman, S.; Nitzan, A. Phenomenology of squeezing and sliding of molecularly thin Xe, CH_4 and $C_{16}H_{34}$ lubrication films between smooth and rough curved solid surfaces with long-range elasticity. *J. Chem. Phys.* **2002**, *117*, 3897–3914. [CrossRef]
41. Plimpton, S. Fast parallel algorithms for short-range molecular dynamics. *J. Comput. Phys.* **1995**, *117*, 1–19. [CrossRef]
42. Mendelev, M.I.; Han, S.; Srolovitz, D.J.; Ackland, G.J.; Sun, D.Y.; Asta, M. Development of new interaction potentials appropriate for crystalline and liquid iron. *Phil. Mag.* **2003**, *83*, 3977–3994. [CrossRef]
43. Martin, M.G.; Siepmann, J.I. Transferable potentials for phase equilibria. 1. United-atom description of n-alkanes. *J. Phys. Chem. B* **1998**, *102*, 2569–2577. [CrossRef]
44. Eggimann, B.L.; Sun, Y.; Dejaco, R.F.; Singh, R.; Ahsan, M.; Josephson, T.R.; Siepmann, J.I. Assessing the quality of molecular simulations for vapor—liquid equilibria: An analysis of the TraPPE database. *J. Chem. Eng. Data* **2020**, *65*, 1330–1344. [CrossRef]
45. Transferable Potentials for Phase Equilibria. Available online: http://trappe.oit.umn.edu/stretch.html (accessed on 15 September 2021).
46. Savio, D.; Fillot, N.; Vergne, P.; Zaccheddu, M. A model for wall slip prediction of confined n-alkanes: Effect of wall-fluid interaction versus fluid resistance. *Tribol. Lett.* **2012**, *46*, 11–22. [CrossRef]
47. Kalyanasundaram, V.; Spearot, D.E.; Malshe, A.P. Molecular dynamics simulation of nanoconfinement induced organization of w-Decane. *Langmuir* **2009**, *25*, 7553–7560. [CrossRef]
48. Evans, D.J.; Holian, B.L. The nose—hoover thermostat. *J. Chem. Phys.* **1985**, *83*, 4069. [CrossRef]
49. Stephan, S.; Dyga, M.; Urbassek, H.M.; Hasse, H. The influence of lubrication and the solid—fluid interaction on thermodynamic properties in a nanoscopic scratching process. *Langmuir* **2019**, *35*, 16948–16960. [CrossRef] [PubMed]
50. An, R.; Huang, L.; Long, Y.; Kalanyan, B.; Lu, X.; Gubbins, K.E. Liquid—solid nanofriction and interfacial wetting. *Langmuir* **2016**, *32*, 743–750. [CrossRef] [PubMed]
51. Tao, X.; Zhao, J.Z.; Xu, K. The ball-bearing effect of diamond nanoparticles as an oil additive. *J. Phys. D Appl. Phys.* **1996**, *29*, 2932–2937. [CrossRef]
52. Peng, D.X.; Chen, C.H.; Kang, Y.; Chang, Y.P.; Chang, S.Y. Size effects of SiO_2 nanoparticles as oil additives on tribology of lubricant. *Ind. Lubr. Tribol.* **2010**, *62*, 111–120. [CrossRef]
53. Sia, S.Y.; Sarhan, A.A.D. Morphology investigation of worn bearing surfaces using SiO_2 nanolubrication system. *Int. J. Adv. Manuf. Technol.* **2014**, *70*, 1063–1071. [CrossRef]
54. Bowden, F.P.; Tabor, D. Friction, lubrication and wear: A survey of work during the last decade. *Br. J. Appl. Phys.* **1966**, *17*, 1521–1544. [CrossRef]
55. Gao, Y.; Brodyanski, A.; Kopnarski, M.; Urbassek, H.M. Nanoscratching of iron: A molecular dynamics study of the influence of surface orientation and scratching direction. *Comput. Mater. Sci.* **2015**, *103*, 77–89. [CrossRef]

metals

MDPI

Article

Concurrent Multiscale Simulations of Rough Lubricated Contact of Aluminum Single Crystal

Jie Zhang [1], Lihong Su [2] and Zhongnan Wang [3,*]

1 State Key Laboratory of Tribology, Tsinghua University, Beijing 100084, China; zj517@mail.tsinghua.edu.cn
2 School of Mechanical, Materials, Mechatronic and Biomedical Engineering, University of Wollongong, Wollongong, NSW 2522, Australia; lihongsu@uow.edu.au
3 School of Mechanical, Electronic and Control Engineering, Beijing Jiaotong University, Beijing 100044, China
* Correspondence: zhn.wang@bjtu.edu.cn; Tel.: +86-010-51683746

Received: 4 June 2020; Accepted: 6 July 2020; Published: 17 July 2020

Abstract: In this paper, a concurrent multiscale simulation strategy coupling atomistic and continuum models was proposed to investigate the three-dimensional contact responses of aluminum single crystal under both dry and lubricated conditions. The Hertz contact is performed by using both the multiscale and full molecular dynamics (MD) simulations for validation. From the contact area, kinetic energy and stress continuity aspects, the multiscale model shows good accuracy. It can also save at least five times the computational time compared with the full MD simulations for the same domain size. Furthermore, the results of lubricated contact show that the lubricant molecules could effectively cover the contact surfaces; thereby separating the aluminum surfaces and bearing the support loads. Moreover, the surface topography could be protected by the thin film formed by the lubricant molecules. It has been found that the contact area decreases obviously with increasing the magnitude of load under both dry and lubricated contacts. Besides, a decrease in contact area is also seen when the number of lubricant molecules increases. The present study has confirmed that the dimension of lubricated contacts could be greatly expanded during the simulation using the proposed multiscale method without sacrificing too much computational time and accuracy.

Keywords: multiscale simulation; surface contact; roughness; mixed lubrication; tribology

---

1. Introduction

Tribology is an interdisciplinary field of research that studies the adhesion, friction, lubrication and wear of contacts from the nano-scale to macro-scale, which is quite complicated, as the processes are usually governed by physical mechanisms at different length scales and can be affected by many factors, such as load, temperature, velocity, environment and so on [1–7]. For contact mechanics, in particular, not only the long-range elastic and plastic deformation play roles, but also the atomic scale roughness at the interface, reported by Luan and Robbins, has critical effects on the contact area and stress [4]. They revealed the breakdown of continuum models, and intrigued new investigations for a number of contacting problems at the nanoscale, which are still open for in-depth analyses. Over the last 40 years, the molecular dynamics (MD) simulations were widely used to explore the mechanisms of contact, friction, wear and lubrication at the nanoscale for various materials [8–12]. As a complement to conventional experiments, MD simulation can produce some detailed observations which cannot be obtained by experimental instruments. Despite the advancing development of computer science, MD still suffers from the temporal and spatial limitations. It is still very challenging to slow the timescale of MD down to realistic vales, although many efforts have been devoted to approaching such a target, such as accelerated molecular dynamics [13] and multiscale simulations [14–16]. The finite element method (FEM), based on the framework of continuum mechanics, has also been widely used to study the elastic and elastic–plastic contact between rough surfaces and deformations [17–20]. However, capturing the accurate information at the atomic scale is difficult by using FEM due to incorrect constitutive law. The development of the coupling methods that bridge the atomistic and continuum models is propelled to overwhelm the limitations of MD and FEM [14–16].

In general, the multiscale model could be divided into three regions: the critical region, the coarse region and the transition region [21–23]. The critical region, composed of atoms, is in charge of the local details of deformation, such as the contact area. The deformation of the coarse region, which is usually far from the critical region, could be described by the continuum models. The region, where the critical and coarse regions overlap, is called the transition region or the handshaking region. It is very critical to complete the information exchange between the two regions, and also to properly remove the unphysically spurious wave at this region [24,25]. Based on such philosophy, several multiscale methods have been established, such as the quasicontinuum method (QCM), bridging scale method (BSM), bridging domain method (BSD) and the hybrid simulation method (HSM). The QCM proposed by Tadmor et al. [21] is one of the pioneering works in multiscale coupling. It has a well-defined energy formulation for both critical and coarse regions. However, the element size in the critical region needs to be refined down to the atomic level. The BSM proposed by Wagner and Liu [22] reduced the atomistic calculations to a small domain with a description of the effects of the eliminated atoms via a time history kernel technique. The BDM proposed by Xiao and Belytschko [23] applied the Lagrange multipliers to impose displacement or velocity compatibility between the atomistic and continuum in a transition domain. For both the BSM and BDM, there is no need to refine the mesh to the atomic size. Particularly, as there is no mathematical framework, it is easier to implement the HSM proposed by Luan et al. [26] than other multiscale methods. Furthermore, the HSM yields the convincing results for the investigations of the contact and friction at the nanoscale benchmarked by pure MD simulations. Therefore, in this study, the HSM was used to study the 3D rough lubricated contact of the single crystal aluminum.

Recently, the HSM has been used for nanoindentation [27] and dry contact between rough surfaces [28,29]. Revealing the behaviors of confined thin lubricant layers between metal surfaces could help understand mixed lubrication. Using the HSM could provide more statistics data and thereby more accurate information compared to the MD simulations [30,31]. Furthermore, the MD region in the HSM can handle the complicated behaviors at the lubricating interfaces, while the coarse region can properly accommodate the elastic deformation from the contact zone [26]. In the previous MD simulations, however, the very limited substrate height with the rigid boundary increased the system rigidity, which could result in the inappropriate accumulation of kinetic energy at the contact zone [32]. To the best of our knowledge, multiscale methods have not been used to perform the simulations of mixed lubrication. In this paper, we used the HSM to study the behaviors of the hexadecane molecules confined within the aluminum tribo-surfaces, which have attracted attention for many years due to their light-weight nature, good formability and low cost [33]. Moreover, Al single crystal as a model material was chosen for comparisons. The changes in the root mean square (RMS) roughness, contact area and surface pressure under dry and lubricated contacts are presented.

2. Methodology

2.1. Model Set-Up

The multiscale (MU) model for the contact between a rigid flat plane and a deformable substrate with a self-affine surface is shown in Figure 1. The substrate was built from Face-center-cubic (FCC) crystal aluminum with a (001) top surface. The lattice constant a of the crystal at 300 K is 0.4045 nm. The substrate dimension is 80a × 80a × 80a, which approximately corresponds to 32.36 nm × 32.36 nm × 32.36 nm. This size is at least 8 times larger than the previous MD simulations [30,34]. Except for the lubricant molecules, the total number of atoms ranges from 670,000 to 700,000 for different surface roughness (0.2 nm, 0.5 nm and 0.8 nm). The specific RMS values were chosen and the rough surfaces were created by using the random midpoint displacement (RMD) algorithm, following the previous MD simulations [34,35]. The corresponding RMS roughness of the generated atomic surface was 0.204, 0.492 and 0.771 nm, respectively. The atomic region of the substrate consists of three layers: (2) the deformable layer, (4) the thermostat layer and (3) the pad layer. The contact occurs among the rigid plane, lubricant and deformable layer. The thermostatting technique was applied to the atoms in the thermostat layer to maintain the constant temperature of 300 K for the system. Along the x and y directions, a periodic boundary condition was applied to both the atomic and FEM regions. The bottom boundary of the FEM region was fixed while the top surface of the deformable layer was free. The FEM region consists of 33,600 elements and 6615 nodes, corresponding to 1,433,600 atoms. A constitutive anisotropic law describes the relation between stress and strain for each integration point. The elastic constants $C_{11} = 114.0$, $C_{12} = 61.6$, $C_{44} = 31.6$ (unit: GPa) were taken from Mishin's work [36].

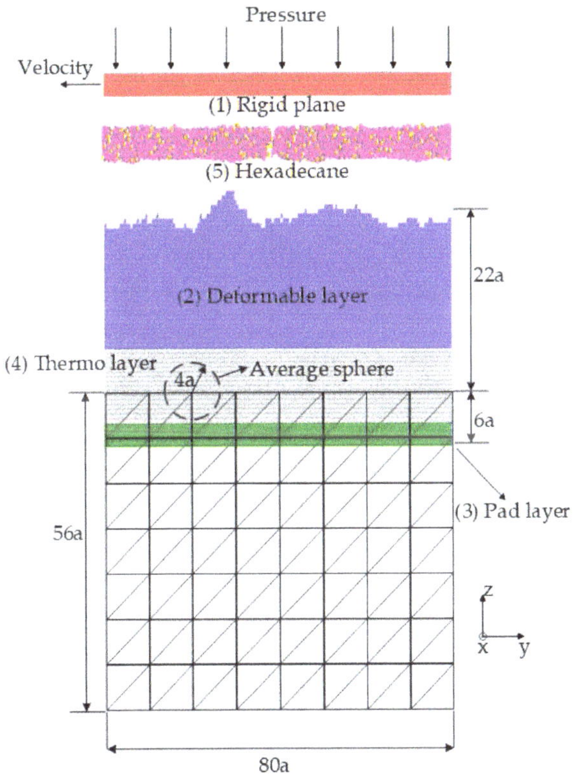

Figure 1. Schematic of the model set-up. a = 0.4045 nm is the lattice constant of the single crystal aluminum.

The key ideas of the coupling strategy for the HSM is described here. The pad layer as a part of the FEM region provides a physical boundary for the atomic region. The displacements of each atom (e.g., $\mathbf{u}_\beta$) in the pad layer are interpolated from the nodal displacements [26] as shown in Equation (1):

$$\mathbf{u}_\beta = \sum_{j=1}^{m_{nodes}} \mathbf{N}_j \mathbf{d}_j, \quad (1)$$

where $\mathbf{N}$ is the shape function of atom β corresponding to the nodes in each element. m_{nodes} is the number of the nodes in each element. As can be seen in Figure 1, the top boundary of the FEM region lies in the thermostat layer. The nodal displacements on this boundary (e.g., $\mathbf{d}_i$) were obtained by averaging the displacements of atoms within the average spheres, as shown in Equation (2). The radius of each sphere is 4a, corresponding to the smallest size of the element edge. The four-node tetrahedron elements were used to divide the continuum region:

$$\mathbf{d}_i = \sum_{\alpha=1}^{n_{atoms}} \omega_\alpha \mathbf{u}_\alpha \tag{2}$$

where n_{atoms} is the number of atoms in the average circle for node i. The weight ω of each atom within the averaging circle with a radius of r_{av} can be calculated by a weighting function [37]:

$$\omega(d) = \begin{cases} 1 - 3d^2 + 2d^3, d < 1 \\ 0, d \geq 1 \end{cases} \tag{3}$$

where $d = r/r_{av}$ and r is the distance between the atom within the average circle and the node.

The timestep is 2 fs. Before loading, the substrate is allowed to relax for equilibrium. The system has 0.4 ns to equilibrate after applying a load. To study the effects of different pressure on the lubricated contact, 0.05, 0.15 and 0.25 GPa were applied on the rigid plane, respectively. During the loading, at an interval of 200 timesteps, the forces and stresses were recorded. The calculation time of the multiscale simulation ranges from 74 to 98 h due to the different number of atoms. For the different cases here, the number of FEM nodes are not changed.

2.2. Force Field

As the plastic deformation on the rough surface with the atoms reshuffling or dislocations usually involves during compression and shearing between the rough surfaces, the EAM potential [38] was used to model the interactions between the aluminum atoms in the substrate. The atomic interactions between the rigid plane and the substrate were calculated by the Lennard–Jones (LJ) potential, which is characterized by the distance and energy parameters σ and ε, respectively. The LJ potential was used to simulate the aluminum oxide surface by reducing ε [34]. A united-atom (UA) model [39] was used to model the hexadecanes molecules. The UA model simplified all the CH_X groups with pseudo carbon atoms. Linear chain molecules of hexadecane with the chemical formula $C_{16}H_{34}$ were chosen as the lubricant liquid. A chain of hexadecane is made of connected CH_3 and CH_2 groups as shown in Figure 2a. The optimization of the conformation of the hexadecane molecule in Figure 2b is attributed to the intra-molecular interactions, which include bond stretching, angle bending and the dihedral angle torsion given in Equations (1)–(3). To study the effects of the amount of lubricant on the lubricated contact, the different number of $C_{16}H_{34}$ molecules was added (0–4000) between the aluminum tribo-surfaces:

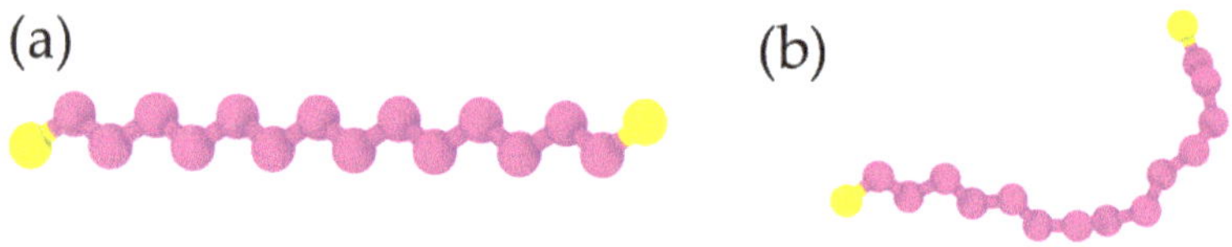

Figure 2. Schematic of one hexadecane molecule's (**a**) initial structure and (**b**) after the optimization of conformation. The pink and yellow atoms indicate the united atoms of CH_2 and CH_3.

$$E_{bond} = k_b(r - r_0)^2, \tag{4}$$

$$E_{bend} = k_\theta(\theta - \theta_0)^2, \tag{5}$$

$$E_{torsion} = c_0 + c_1[1 + \cos\phi] + c_2[1 - \cos(2\phi)] + c_3[1 + \cos(3\phi)], \tag{6}$$

The corresponding parameters in Equations (4)–(6) are taken from Reference [39]. The inter-molecular and liquid–solid interactions are described by the LJ potential with a cutoff distance of 1.4 nm. The LJ parameters for the liquid–solid interactions were estimated via the Lorentz–Berthelot mixing rules (Table 1).

Table 1. Parameters for the Lennard–Jones (LJ) potentials for the non-bonded interactions.

Neighbors	σ (nm)	ε (eV)
Al–Al [a]	0.2596	0.010677
Al–CH_2 [c]	0.3273	0.006506
AlCH_3 [c]	0.3173	0.009496
CH_2–CH_2 [b]	0.3950	0.003964
CH_3–CH_3 [b]	0.3750	0.008445
CH_3–CH_2 [c]	0.3850	0.005786

[a] taken from Ref. [34]. [b] taken from Ref. [39]. [c] calculated from the Lorentz–Berthelot rules.

2.3. Contact Area

Calculations for the real contact area are of importance at the atomic scale, especially when the dimension of the devices is reduced down to the nanometer. The real contact area has an important effect on heat conduction, friction and adhesion. In continuum mechanics, the contact area can be obtained based on the edge of the contact zone. It is difficult to accurately define the atomic area and even the atomic contact. The atomic contact is determined by either repulsive force [28,40] or distance [35,41]. Here, the contact is established when the distance of two atoms is less than or equals to 0.5 nm [30], and the real contact area is estimated from a projected rectangular cuboid method [34]. In this method, the n × n grids as a bottom face of the rectangular cuboids are plotted on the x–y plane, on which all the atoms of the deformable layer are projected. Those atoms are divided into the corresponding grids according to their x and y coordinates. The length of each cuboid is the maximum z coordinate values of the atoms within it. Similar cuboids can be plotted for the rigid plane. The top faces of the deformable layer cuboids and the bottom faces of the rigid plane cuboids are used to determine the atomic contact.

2.4. Multiscale Strategy Valication

Here, we validate our MU model by performing a 3D Hertz contact, benchmarked with full MD simulations, following the the Anciaux and Molinali's work [32]. Details of the model setup for the Hertz contact is shown in Figure S1 and demonstrated in the supplementary materials. The contact area, kinetic energy and stress continuality were considered. Figure S2a shows the relationship between the contact area and the normalized indent displacement. The contact region consists of the atoms on the stepped tip and is close to the area of a circle. The contact area was determined by averaging the distances between the center of the circle and the outermost atoms. In terms of the contact area, the MU predicts the MD well. It is found from Figure S2a that the results between the two models are very close. From the energy point of view, the MU works well based on the results in Figure S2b. Although there are a few deviating points, the kinetic energies of MU and MD are close to each other. In the MU model, the atomic, or the virial stress was calculated for the MD region while the Cauchy stress for the FEM region. The stress component σ_{zz} of both the MU model and the MD model when normalized indent displacement is 1.31a was shown in Figure S2c. First and foremost, the HSM can ensure stress continuity after comparing σ_{zz} of both the FEM region and the MD region (dotted square). Another important result shown in Figure S2c is that the MU model can predict the full MD model well.

From the contact area, kinetic energy and stress continuity aspects, our MU simulations show good accuracy compared with the full MD simulations. More importantly, it can save at least 5 times the computational time compared with the full MD simulations for the same domain size. With relatively good computational cost and accuracy, the MU model was here extended to study lubricated contacts.

3. Results and Discussion

3.1. RMS Roughness

Before compression, the rigid plane was above the substrate with a distance larger than the cut-off radius of the interaction potential. As mentioned above, the original RMS roughness of the generated atomic surface was 0.204, 0.492 and 0.771 nm, respectively. The atoms at the surface lack the full neighbors, and are attracted by the atoms at the subsurface. Therefore, the relaxing process at a temperature of 300 K modified the surface roughness with RMS = 0.198, 0.484 and 0.760 nm, respectively. Although the RMS roughness of the surfaces changes, it is still close to the required values. After relaxation, the top plane was forced to move downwards under constant pressure. Without any resisting force, the top plane will impact on the substrate with a high speed. This is not acceptable, since it will cause significant plastic deformation on the substrate. Therefore, the top plane was forced to move at a very low speed until it touched the deformable layer. The system heights were calculated during the whole compression. In Figure 3, the relationship between the system height and simulation time for dry and lubricated contact is shown. It indicates that all the simulations reached equilibrium after 0.4 ns. The heights in the lubricated cases were higher than the dry counterparts, as shown in Figure 3a,b. The thickness of the thin film formed by the hexadecane molecules contributed to the system height. In Figure 3c, two thousand hexadecane molecules had no effect on the system height when the surface roughness was 0.8 nm (indicated by the dashed circle in the Figure), since there was not enough quantity to fill the cavity between the two contacting surfaces. Meanwhile, four thousand hexadecane molecules lifted the top plane by a distance of 0.4 nm as indicated by the dashed arrow in Figure 3c. The system height is also related to the chain length of the lubricant molecules [42]. The effective separation for the contacting surfaces is dependent on the added amount of lubricant and surface roughness. A few lubricant molecules can work well for relatively flat surfaces, while much more molecules are required to fill a cavity in rougher surfaces and effectively separate them. The volume of the hexadecane molecule is also related to the rotation of C–C bonds—thereby affecting the system heights in Figure 3.

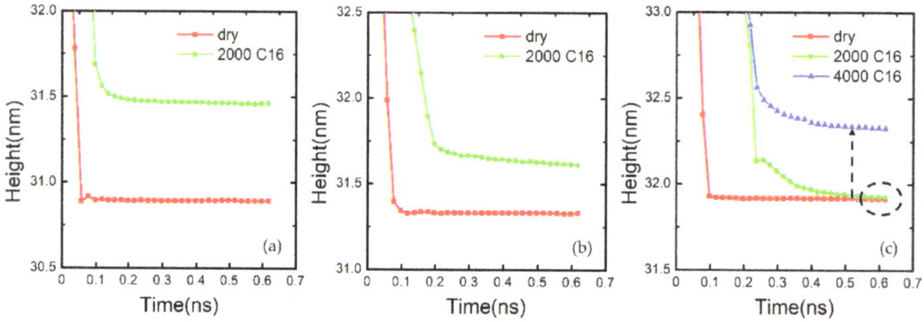

Figure 3. Variation of the system heights with the time for the root mean square (RMS) = (**a**) 0.2 nm, (**b**) 0.5 nm, (**c**) 0.8 nm under 0.25 MPa. The red lines indicate dry contact while the green and blue lines indicate lubricated contact with 2000 and 4000 hexadecane molecules, respectively.

When the contact occurs, the top plane will flatten the substrate asperities as shown in Figure 4b,c. The RMS roughness of flattened surface changes as the load increases. The RMS of the flattened surface is denoted as RMS_f, while that of the initial surface RMS_u. The change of RMS, therefore, is defined as $\Delta RMS = RMS_f - RMS_u$. Firstly, it indicates that RMS_f decreases with the increasing loads for three surfaces, and this relationship is independent of RMS_u. However, the percentage of change in RMS roughness is related to RMS_u. There was a much smaller change for the rougher surfaces (solid triangle and square in Figure 4a), because the contact for those rougher surfaces occurred on a few high asperities which support the loads. It is the rest part on the surface that dominated the roughness. However, the surface RMS changed more for RMS = 0.2 nm, because much more asperities on the substrate contacted with the top plane were simultaneously flattened. The peak-to-valley (PV) distances [34] indicate roughly the number of the contacting asperities. The PV distance here is determined by the difference between the maximum peak height to the maximum valley depth of the atomistic surface. The PV of the atomic surfaces for RMS = 0.2, 0.5 and 0.8 nm are 0.172, 0.3285 and 0.514 nm, respectively. The height of the highest asperity for RMS = 0.2 nm is approximately 0.86 nm. Therefore, many asperities support the top plane simultaneously. Figure 5 shows the effects of the molecule number on the RMS change. Zheng et al. [42] demonstrated that the appropriate amount of lubricant is dependent on the RMS roughness. The increase in the molecule number reduces the change in RMS roughness or asperity flattening as shown in Figure 5. After compression, the lubricant molecules filled the valleys and covered the surface as shown in the insets. With 4000 molecules, the surface could be entirely covered. The asperities were covered by the lubricant, and were not contacted by the top plane. Consequently, the RMS changed only by about 1.8%. In addition, Wu et al. [43] indicated that the lubricant viscosity has an important effect on the changes in the surface roughness. In their work, the smaller reduction in the surface roughness was obtained for the lubricant with a higher viscosity. The viscosity in Wu's work plays a similar role as the lubricant amount here. A well-formed lubricant film reduced the direct metal–metal contact, and protected the surfaces. For the lubricated contact, the new surface was formed by the lubricant molecules and substrate atoms. The RMS roughness of the new surface was 0.33 and 0.06 nm for 3000 and 4000 molecules, respectively. They are both much lower than the original roughness of 0.8 nm.

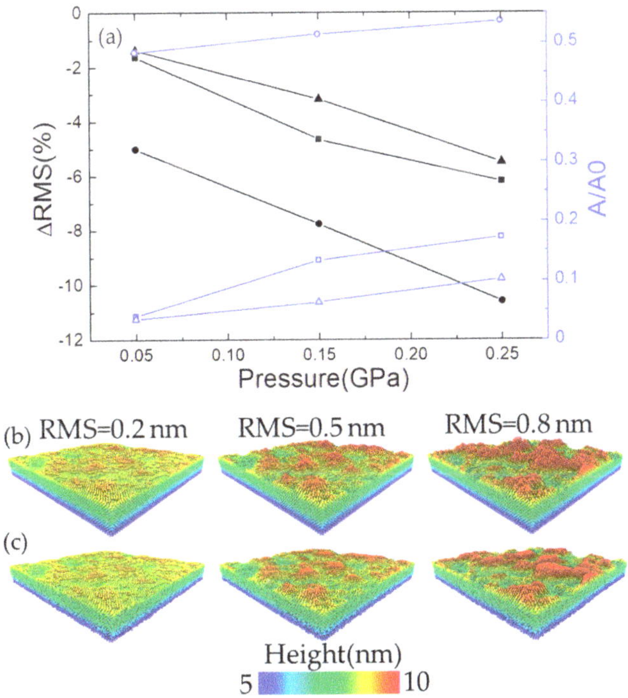

Figure 4. (**a**) RMS change ΔRMS and contact area A/A0 of dry contact as a function of pressure. The circle, square and triangle indicate RMS = 0.2, 0.5 and 0.8 nm, respectively. Changes in the topography before (**b**) and after (**c**) loading for the surfaces with RMS = 0.2, 0.5 and 0.8 nm, respectively.

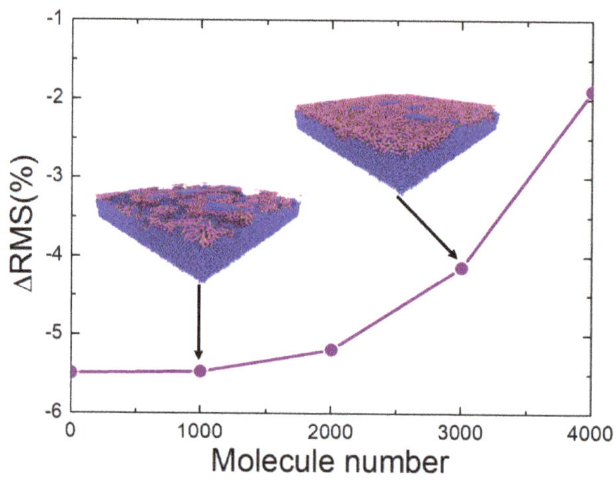

Figure 5. Variations of the RMS change ΔRMS with the molecule number under 0.25 GPa. The RMS roughness of the surface is 0.8 nm. Insets show the surface topography and the hexadecane molecule distribution after loading.

3.2. Contact Area

The effects of roughness and lubricant on the contact area calculated by the project method will be studied in this section. Figure 4a indicates that the contact area goes up with the increasing loads for the surfaces with three different RMS. Under the load of 0.05 GPa, the contact area of RMS = 0.2 nm is close to 50%, which is much higher than 4% and 3.5% of the RMS = 0.5 and 0.8 nm. As mentioned in the previous section, the heights of the asperities on the surface of RMS = 0.2 nm are relatively low, so a number of them were in contact simultaneously. For the rougher surface, however, only few asperities supported the load applied from the top plane. With the increasing loads, the contacting asperity deformed and their heights decreased while the lower non-contacting asperities were joined to support the top plane. Therefore, the contact area of the three surfaces was increased in Figure 4a. Considering the different topography due to the different random seed when generating the self-affine surfaces [35] and adhesive strength [34], the results could deviate a little. The relationship between the contact area and the loads obtained in this work qualitatively agrees with the published literature [44].

For the lubricated contacts, the contact area still increased with the loads at the range of 0.05 GPa (red diamond) to 0.25 GPa (purple square) as shown in Figure 6a. However, the contact area reduced with the increasing number of molecules. The lubricants could separate the substrate from the top plane. The separating gap depends on the lubricant amount. It indicates that under 0.25 GPa the contact area of 0 and 1000 molecules is almost identical while the contact area of 4000 molecules was reduced by 75%. However, it was not enough for the 2000 molecules to fill the valleys on the rough surface. From the point of view of normal force, it could also provide some clues. The normal force on the top plane of lubricated contact was divided into two parts: normal force F_S from the substrate and normal force F_L from the lubricant. The sum of F_S and F_L is equal to the normal force F_0 on the top plane of dry contact (no lubricant molecules). Figure 6b shows the normal force on the top plane supported by dry and lubricated contacts as a function of molecule number. In the figure, the support force F_S/F_0 from the asperities roughly decreased with the larger number of molecules, which is consistent with the trend in the contact area. At the same time, the normal force F_L/F_0 from the lubricant increased with the higher number of molecules. The decrease in the normal force F_S/F_0 under each load indicates that the number of the contacting asperities dropped. At 0.25 GPa, more lubricant endured the larger force. When 4000 hexadecane molecules were used, F_L reached 87.17 nN under

0.25 GPa, which corresponds to 85.45% of F_S in dry contact. The project method can be used to calculate the volume occupied by the hexadecane molecules—thereby estimating its density. For RMS = 0.8 nm with 4000 molecules under 0.15 and 0.25 GPa, the hexadecane has the density of 6.9 and 7.2 g/cm^3, respectively. The estimations are very close to 7.7 g/cm^3 obtained from the previous experiments [45].

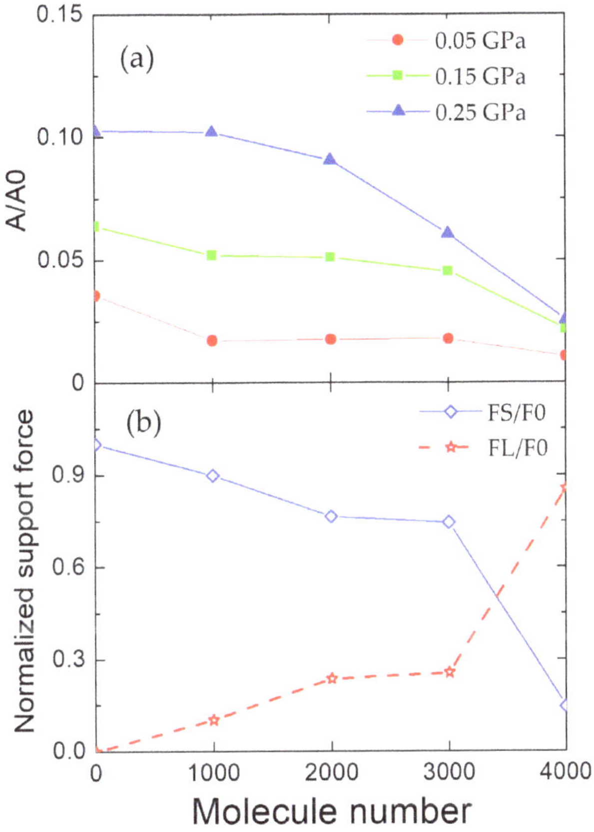

Figure 6. Contact area (**a**) and the normalized normal force (**b**) as a function of the lubricant molecule number for surface with a roughness RMS = 0.8 nm. The molecule number of zero indicates dry contact. The load is 0.25 GPa in (**b**).

3.3. Surface Pressure

Surface pressure was calculated based on the projection method in Section 2.3. There are other approaches for calculating surface pressure, but for the comparisons the projection method was used. The virial stress of each atom was calculated at every 20 ps. After the cuboids were obtained, the surface pressure was estimated by averaging the virial stress along the z direction of all atoms within each cuboid. Before that, we show in Figure 7 the deformation of the FEM region due to the applied loads from the top plane on the substrate. Before compression, there is almost no deformation at the coarse region as shown in Figure 7a. With the compression of the top plane, the deformation is small in this region as shown in Figure 7b, as the FEM region is far from the surface. However, more deformation occurred in the FEM region, since the asperities were flattened in the corresponding area on the surface, indicating that the coarse region can properly accommodate the elastic deformation from the contact zone.

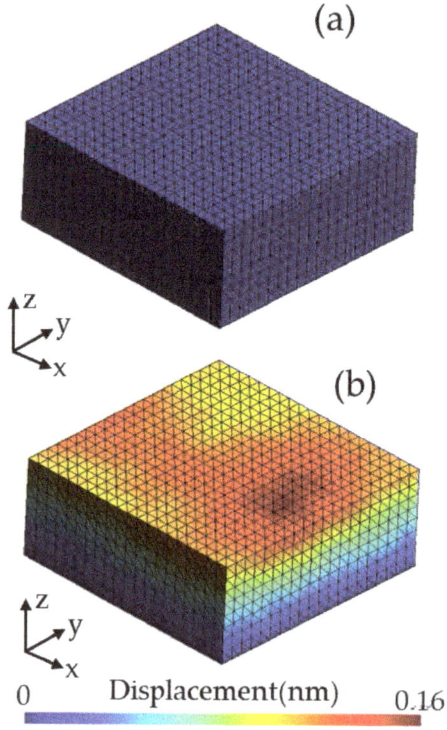

Figure 7. Deformation of the FEM region (**a**) before compression and (**b**) after compression for dry contact.

Figure 8 shows the pressure distribution for the unlubricated and lubricated contacts. To obtain the distribution histogram, the surface pressure in each cuboid was binned with an interval of 0.05 GPa. The shape of the probability distribution in this work was in agreement with the previous works [30,34]. It seems that the system size and the roughness of the top surface have little effect on the shape of the probability distribution. In Figure 8a, the sum of the negative pressure distribution is 88.1%, 80.0% and 66.79% for 0.05, 0.15 and 0.25 GPa, respectively. Meanwhile, in Figure 8b, the obtained values are 89.2%, 72.8% and 63.4%, respectively. More than half of the surface area mainly suffered the attractive forces. In addition, for both dry and lubricated surfaces, the probability sum of the positive pressure goes up as the loads increase as shown in Figure 8. Further compression caused the increase in the repulsive forces between the top plane, lubricants and substrate. As demonstrated by Cheng et al. [46], the lubricant molecules spread the pressure over a larger contact area. The shifting of probability with the increasing load for the lubricated contacts is much more than that for the dry contacts, indicating the significant role of lubricant molecules in bearing the loads.

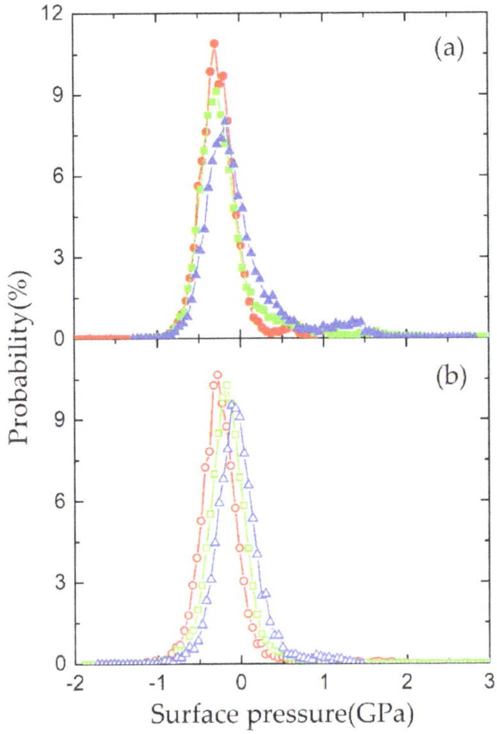

Figure 8. Pressure distribution in the dry contact (**a**) and 4000 molecules lubricated contact (**b**) under 0.05 GPa (red line), 0.15 GPa (green line) and 0.25 GPa (blue line).

4. Conclusions

In the present study, a concurrent multiscale model that combines the MD and FEM methods was proposed and successfully applied to study the three-dimensional rough lubricated contact of aluminum single crystal. We performed the Hertz contact to compare the MU model with the full MD model. Our MU simulations show good accuracy from the contact area, kinetic energy and stress continuity aspects, and it can save at least five times the computational time compared with the full MD simulations for the same domain size. With relatively good computational cost and accuracy, the MU model here has been extended to study lubricated contacts. A systematic investigation considering the effects of the lubricant amount, surface roughness and loads during contact in mixed lubrication at the atomic scale was conducted. It was found that the sufficient lubricant molecules could fill the cavity and cover the large portion of contacting surfaces and effectively separate the tribo-pair. Meanwhile, the insufficient lubricant molecules can work well for relatively flat surfaces. In addition, the results show that the increasing number of lubricant molecules leads to smaller changes in surface RMS under the studied loads, varying from 0.05 to 0.25 GPa, indicating that the thin film formed by the lubricant molecules is beneficial to maintain the initial surface topography. For both the lubricated and unlubricated contacts, the contact area decreases with the increasing loads. As the tribo-pair gap increases with the increasing lubricant amount, the contact area decreases, and the larger portion of loads is supported by the lubricant molecules. The changes in pressure distribution for the different loads in the lubricated contacts confirm the contributions of the lubricant molecules to the load-bearing.

Supplementary Materials: The following are available online at http://www.mdpi.com/2075-4701/10/7/965/s1, Figure S1: Geometry of the multiscale (a) and full MD (b) simulations, Figure S2: Comparisons between MU and full MD simulations: (a) Normalized contact area as a function of normalized indent displacement. (b) Kinetic energy near the substrate surface as a function normalized indent displacement. (c) Changes of stress component σ_{zz} in a line along the z direction when normalized indent displacement is 1.31a. Indent displacement and Z are normalized by lattice constant of a.

Author Contributions: Conceptualization, software, visualization, writing—original draft, J.Z.; methodology, J.Z. and L.S.; validation, writing—review and editing, L.S. and Z.W. All authors have read and agreed to the published version of the manuscript.

Funding: This research was funded by the Fundamental Research Funds for the Central Universities (grant number 2020JBM059), the National Natural Science Foundation of China (grant number 51905296) and Australian Research Council (ARC) Discovery Early Career Researcher Award (DECRA) project (DE180100124).

Conflicts of Interest: The authors declare no conflict of interest.

References

1. Persson, B.N.J.; Albohr, O.; Tartaglino, U.; Volokitin, A.I.; Tosatti, E. On the nature of surface roughness with application to contact mechanics, sealing, rubber friction and adhesion. *J. Phys. Condens. Matter* **2004**, *17*, R1–R62. [CrossRef] [PubMed]
2. Bhushan, B. Contact mechanics of rough surfaces in tribology: Multiple asperity contact. *Tribol. Lett.* **1998**, *4*, 1–35. [CrossRef]
3. Manini, N.; Mistura, G.; Paolicelli, G.; Tosatti, E.; Vanossi, A. Current trends in the physics of nanoscale friction. *Adv. Phys. X* **2017**, *2*, 569–590. [CrossRef]
4. Luan, B.; Robbins, M.O. The breakdown of continuum models for mechanical contacts. *Nature* **2005**, *435*, 929–932. [CrossRef]
5. Deng, G.; Tieu, A.K.; Lan, X.; Su, L.; Wang, L.; Zhu, Q.; Zhu, H. Effects of normal load and velocity on the dry sliding tribological behaviour of CoCrFeNiMo0.2 high entropy alloy. *Tribol. Int.* **2020**, *144*, 106116. [CrossRef]
6. Ta, T.D.; Tieu, A.K.; Zhu, H.; Zhu, Q.; Kosasih, P.B.; Zhang, J.; Deng, G. Tribological Behavior of Aqueous Copolymer Lubricant in Mixed Lubrication Regime. *ACS Appl. Mater. Interfaces* **2016**, *8*, 5641–5652. [CrossRef]
7. Deng, G.Y.; Tieu, A.K.; Su, L.H.; Zhu, H.T.; Reid, M.; Zhu, Q.; Kong, C. Microstructural study and residual stress measurement of a hot rolling work roll material during isothermal oxidation. *Int. J. Adv. Manuf. Technol.* **2019**, *102*, 2107–2118. [CrossRef]
8. Zhang, J.; Xu, Q.; Gao, L.; Ma, T.; Qiu, M.; Hu, Y.; Wang, H.; Luo, J. A molecular dynamics study of lubricating mechanism of graphene nanoflakes embedded in Cu-based nanocomposite. *Appl. Surf. Sci.* **2020**, *511*, 145620. [CrossRef]
9. Ma, T.-B.; Hu, Y.-Z.; Wang, H. Molecular dynamics simulation of shear-induced graphitization of amorphous carbon films. *Carbon* **2009**, *47*, 1953–1957. [CrossRef]
10. Kong, N.; Wei, B.; Zhuang, Y.; Zhang, J.; Li, H.; Wang, B. Effect of Compressive Prestrain on the Anti-Pressure and Anti-Wear Performance of Monolayer MoS2: A Molecular Dynamics Study. *Nanomaterials* **2020**, *10*, 275. [CrossRef]
11. Zhao, X.; Lu, C.; Tieu, A.K.; Zhan, L.; Huang, M.; Su, L.; Pei, L.; Zhang, L. Deformation twinning and dislocation processes in nanotwinned copper by molecular dynamics simulations. *Comp. Mater. Sci.* **2018**, *142*, 59–71. [CrossRef]
12. Zhang, L.; Shibuta, Y.; Lu, C.; Huang, X. Interaction between nano-voids and migrating grain boundary by molecular dynamics simulation. *Acta Mater.* **2019**, *173*, 206–224. [CrossRef]
13. Li, Q.; Dong, Y.; Perez, D.; Martini, A.; Carpick, R.W. Speed Dependence of Atomic Stick-Slip Friction in Optimally Matched Experiments and Molecular Dynamics Simulations. *Phys. Rev. Lett.* **2011**, *106*, 126101. [CrossRef] [PubMed]
14. Zhang, J.; Michal, G.; Tieu, A.K.; Zhu, H.T.; Deng, G.Y. Hertz Contact at the Nanoscale with a 3D Multiscale Model. *Appl. Mech. Mater.* **2016**, *846*, 306–311. [CrossRef]
15. Liu, W.K.; Karpov, E.G.; Zhang, S.; Park, H.S. An introduction to computational nanomechanics and materials. *Comput. Method Appl. Mech. Eng.* **2004**, *193*, 1529–1578. [CrossRef]

16. Michal, G.; Lu, C.; Kiet, T.A. Multiscale model of elastic nanocontacts. *Comp. Mater. Sci.* **2014**, *81*, 98–103. [CrossRef]
17. Hyun, S.; Pei, L.; Molinari, J.F.; Robbins, M.O. Finite-Element analysis of contact between elastic self-affine surfaces. *Phys. Rev. E* **2004**, *70*, 026117. [CrossRef]
18. Pei, L.; Hyun, S.; Molinari, J.F.; Robbins, M.O. Finite element modeling of elasto-Plastic contact between rough surfaces. *J. Mech. Phys. Solids* **2005**, *53*, 2385–2409. [CrossRef]
19. Deng, G.Y.; Tieu, A.K.; Su, L.H.; Zhu, H.T.; Zhu, Q.; Zamri, W.F.H.; Kong, C. Characterizing deformation behaviour of an oxidized high speed steel: Effects of nanoindentation depth, friction and oxide scale porosity. *Int. J. Mech. Sci.* **2019**, *155*, 267–285. [CrossRef]
20. Deng, G.Y.; Zhu, H.T.; Tieu, A.K.; Su, L.H.; Reid, M.; Zhang, L.; Wei, P.T.; Zhao, X.; Wang, H.; Zhang, J.; et al. Theoretical and experimental investigation of thermal and oxidation behaviours of a high speed steel work roll during hot rolling. *Int. J. Mech. Sci.* **2017**, *131–132*, 811–826. [CrossRef]
21. Tadmor, E.B.; Ortiz, M.; Phillips, R. Quasicontinuum analysis of defects in solids. *Philos. Mag. A* **1996**, *73*, 1529–1563. [CrossRef]
22. Wagner, G.J.; Liu, W.K. Coupling of atomistic and continuum simulations using a bridging scale decomposition. *J. Comput. Phys.* **2003**, *190*, 249–274. [CrossRef]
23. Xiao, S.P.; Belytschko, T. A bridging domain method for coupling continua with molecular dynamics. *Comput. Method Appl. Mech. Eng.* **2004**, *193*, 1645–1669. [CrossRef]
24. Park, H.S.; Liu, W.K. An introduction and tutorial on multiple-scale analysis in solids. *Comput. Method Appl. Mech. Eng.* **2004**, *193*, 1733–1772. [CrossRef]
25. Zhang, J.; Tieu, K.; Michal, G.; Zhu, H.; Zhang, L.; Su, L.; Deng, G.; Wang, H. A damping boundary condition for atomistic-continuum coupling. *Chin. Phys. B* **2017**, *26*, 68702. [CrossRef]
26. Luan, B.Q.; Hyun, S.; Molinari, J.F.; Bernstein, N.; Robbins, M.O. Multiscale modeling of two-Dimensional contacts. *Phys. Rev. E* **2006**, *74*, 046710. [CrossRef]
27. Zhu, P.; Hu, Y.; Fang, F.; Wang, H. Multiscale simulations of nanoindentation and nanoscratch of single crystal copper. *Appl. Surf. Sci.* **2012**, *258*, 4624–4631. [CrossRef]
28. Luan, B.; Robbins, M.O. Hybrid Atomistic/Continuum Study of Contact and Friction between Rough Solids. *Tribol. Lett.* **2009**, *36*, 1–16. [CrossRef]
29. Zhang, J.; Zhang, L.; Tieu, A.K.; Michal, G.; Zhu, H.T.; Deng, G.Y. Finite-Temperature Multiscale Simulations for 3D Nanoscale Contacts. *Appl. Mech. Mater.* **2016**, *846*, 288–293. [CrossRef]
30. Zheng, X.; Zhu, H.; Kiet, T.A.; Kosasih, B. A molecular dynamics simulation of 3D rough lubricated contact. *Tribol. Int.* **2013**, *67*, 217–221. [CrossRef]
31. Zheng, X.; Zhu, H.; Tieu, A.K.; Kosasih, B. Roughness and Lubricant Effect on 3D Atomic Asperity Contact. *Tribol. Lett.* **2013**, *53*, 215–223. [CrossRef]
32. Anciaux, G.; Molinari, J.-F. Contact mechanics at the nanoscale, a 3D multiscale approach. *Int. J. Numer. Methods Eng.* **2009**, *79*, 1041–1067. [CrossRef]
33. Su, L.; Deng, G.; Luzin, V.; Wang, H.; Wang, Z.; Yu, H.; Li, H.; Tieu, A.K. Effect of cryogenic temperature equal channel angular pressing on microstructure, bulk texture and tensile properties of AA1050. *Mater. Sci. Eng. A* **2020**, *780*, 139190. [CrossRef]
34. Spijker, P.; Anciaux, G.; Molinari, J.-F. The effect of loading on surface roughness at the atomistic level. *Comput. Mech.* **2012**, *50*, 273–283. [CrossRef]
35. Spijker, P.; Anciaux, G.; Molinari, J.-F. Relations between roughness, temperature and dry sliding friction at the atomic scale. *Tribol. Int.* **2013**, *59*, 222–229. [CrossRef]
36. Mishin, Y.; Farkas, D.; Mehl, M.J.; Papaconstantopoulos, D.A. Interatomic potentials for monoatomic metals from experimental data and ab initio calculations. *Phys. Rev. B* **1999**, *59*, 3393–3407. [CrossRef]
37. Miller, R.E.; Tadmor, E.B. A unified framework and performance benchmark of fourteen multiscale atomistic/continuum coupling methods. *Model. Simul. Mater. Sci. Eng.* **2009**, *17*, 053001. [CrossRef]
38. Foiles, S.M.; Baskes, M.I.; Daw, M.S. Embedded-atom-method functions for the fcc metals Cu, Ag, Au, Ni, Pd, Pt, and their alloys. *Phys. Rev. B* **1986**, *33*, 7983–7991. [CrossRef]
39. Martin, M.G.; Siepmann, J.I. Transferable Potentials for Phase Equilibria. 1. United-Atom Description of n-Alkanes. *J. Phys. Chem. B* **1998**, *102*, 2569–2577. [CrossRef]
40. Mo, Y.; Turner, K.T.; Szlufarska, I. Friction laws at the nanoscale. *Nature* **2009**, *457*, 1116–1119. [CrossRef]

41. Anciaux, G.; Ramisetti, S.B.; Molinari, J.F. A finite temperature bridging domain method for MD-FE coupling and application to a contact problem. *Comput. Method Appl. Mech. Eng.* **2012**, *205–208*, 204–212. [CrossRef]
42. Zheng, X.; Zhu, H.; Kosasih, B.; Tieu, A.K. A molecular dynamics simulation of boundary lubrication: The effect of n-alkanes chain length and normal load. *Wear* **2013**, *301*, 62–69. [CrossRef]
43. Wu, C.; Zhang, L.; Li, S.; Jiang, Z.; Qu, P. A unified method for characterizing multiple lubrication regimes involving plastic deformation of surface asperities. *Tribol. Int.* **2016**, *100*, 70–83. [CrossRef]
44. Campañá, C.; Müser, M.H. Contact mechanics of real vs. randomly rough surfaces: A Green's function molecular dynamics study. *Europhys. Lett.* **2007**, *77*, 38005. [CrossRef]
45. Tanaka, H.; Yamaki, Y.; Kato, M. Solubility of carbon dioxide in pentadecane, hexadecane, and pentadecane + hexadecane. *J. Chem. Eng. Data* **1993**, *38*, 386–388. [CrossRef]
46. Cheng, S.; Luan, B.; Robbins, M.O. Contact and friction of nanoasperities: Effects of adsorbed monolayers. *Phys. Rev. E* **2010**, *81*, 016102. [CrossRef] [PubMed]

MDPI
St. Alban-Anlage 66
4052 Basel
Switzerland
Tel. +41 61 683 77 34
Fax +41 61 302 89 18
www.mdpi.com

Metals Editorial Office
E-mail: metals@mdpi.com
www.mdpi.com/journal/metals

www.ingramcontent.com/pod-product-compliance
Lightning Source LLC
Chambersburg PA
CBHW061620210926
51850CB00009B/2382
9783036573434